Research Notes in Mathematics

Submission of proposals for consideration
Suggestions for publication, in the form of outlines and representative samples, are invited by the editorial board for assessment. Intending authors should contact either the main editor or another member of the editorial board, citing the relevant AMS subject classifications. Refereeing is by members of the board and other mathematical authorities in the topic concerned, located throughout the world.

Preparation of accepted manuscripts
On acceptance of a proposal, the publisher will supply full instructions for the preparation of manuscripts in a form suitable for direct photo-lithographic reproduction. Specially printed grid sheets are provided and a contribution is offered by the publisher towards the cost of typing.

Illustrations should be prepared by the authors, ready for direct reproduction without further improvement. The use of hand-drawn symbols should be avoided wherever possible, in order to maintain maximum clarity of the text.

The publisher will be pleased to give any guidance necessary during the preparation of a typescript, and will be happy to answer any queries.

Important note
In order to avoid later retyping, intending authors are strongly urged not to begin final preparation of a typescript before receiving the publisher's guidelines and special paper. In this way it is hoped to preserve the uniform appearance of the series.

Titles in this series

1 Improperly posed boundary value problems
A Carasso and A P Stone

2 Lie algebras generated by finite dimensional ideals
I N Stewart

3 Bifurcation problems in nonlinear elasticity
R W Dickey

4 Partial differential equations in the complex domain
D L Colton

5 Quasilinear hyperbolic systems and waves
A Jeffrey

6 Solution of boundary value problems by the method of integral operators
D L Colton

7 Taylor expansions and catastrophes
T Poston and I N Stewart

8 Function theoretic methods in differential equations
R P Gilbert and R J Weinacht

9 Differential topology with a view to applications
D R J Chillingworth

10 Characteristic classes of foliations
H V Pittie

11 Stochastic integration and generalized martingales
A U Kussmaul

12 Zeta-functions: An introduction to algebraic geometry
A D Thomas

13 Explicit *a priori* inequalities with applications to boundary value problems
V G Sigillito

14 Nonlinear diffusion
W E Fitzgibbon III and H F Walker

15 Unsolved problems concerning lattice points
J Hammer

16 Edge-colourings of graphs
S Fiorini and R J Wilson

17 Nonlinear analysis and mechanics: Heriot-Watt Symposium Volume I
R J Knops

18 Actions of fine abelian groups
C Kosniowski

19 Closed graph theorems and webbed spaces
M De Wilde

20 Singular perturbation techniques applied to integro-differential equations
H Grabmüller

21 Retarded functional differential equations: A global point of view
S E A Mohammed

22 Multiparameter spectral theory in Hilbert space
B D Sleeman

24 Mathematical modelling techniques
R Aris

25 Singular points of smooth mappings
C G Gibson

26 Nonlinear evolution equations solvable by the spectral transform
F Calogero

27 Nonlinear analysis and mechanics: Heriot-Watt Symposium Volume II
R J Knops

28 Constructive functional analysis
D S Bridges

29 Elongational flows: Aspects of the behaviour of model elasticoviscous fluids
C J S Petrie

30 Nonlinear analysis and mechanics: Heriot-Watt Symposium Volume III
R J Knops

31 Fractional calculus and integral transforms of generalized functions
A C McBride

32 Complex manifold techniques in theoretical physics
D E Lerner and P D Sommers

33 Hilbert's third problem: scissors congruence
C-H Sah

34 Graph theory and combinatorics
R J Wilson

35 The Tricomi equation with applications to the theory of plane transonic flow
A R Manwell

36 Abstract differential equations
S D Zaidman

37 Advances in twistor theory
L P Hughston and R S Ward

38 Operator theory and functional analysis
I Erdelyi

39 Nonlinear analysis and mechanics: Heriot-Watt Symposium Volume IV
R J Knops

40 Singular systems of differential equations
S L Campbell

41 N-dimensional crystallography
R L E Schwarzenberger

42 Nonlinear partial differential equations in physical problems
D Graffi

43 Shifts and periodicity for right invertible operators
D Przeworska-Rolewicz

44 Rings with chain conditions
A W Chatters and C R Hajarnavis

45 Moduli, deformations and classifications of compact complex manifolds
D Sundararaman

46 Nonlinear problems of analysis in geometry and mechanics
M Atteia, D Bancel and I Gumowski

47 Algorithmic methods in optimal control
W A Gruver and E Sachs

48 Abstract Cauchy problems and functional differential equations
F Kappel and W Schappacher

49 Sequence spaces
W H Ruckle

50 Recent contributions to nonlinear partial differential equations
H Berestycki and H Brezis

51 Subnormal operators
J B Conway

52 Wave propagation in viscoelastic media
F Mainardi

53 Nonlinear partial differential equations and their applications: Collège de France Seminar. Volume I
H Brezis and J L Lions

54 Geometry of Coxeter groups
H Hiller

55 Cusps of Gauss mappings
T Banchoff, T Gaffney and C McCrory

56 An approach to algebraic K-theory
A J Berrick
57 Convex analysis and optimization
J-P Aubin and R B Vintner
58 Convex analysis with applications in the differentiation of convex functions
J R Giles
59 Weak and variational methods for moving boundary problems
C M Elliott and J R Ockendon
60 Nonlinear partial differential equations and their applications: Collège de France Seminar. Volume II
H Brezis and J L Lions
61 Singular systems of differential equations II
S L Campbell
62 Rates of convergence in the central limit theorem
Peter Hall
63 Solution of differential equations by means of one-parameter groups
J M Hill
64 Hankel operators on Hilbert space
S C Power
65 Schrödinger-type operators with continuous spectra
M S P Eastham and H Kalf
66 Recent applications of generalized inverses
S L Campbell
67 Riesz and Fredholm theory in Banach algebra
B A Barnes, G J Murphy, M R F Smyth and T T West
68 Evolution equations and their applications
F Kappel and W Schappacher
69 Generalized solutions of Hamilton-Jacobi equations
P L Lions
70 Nonlinear partial differential equations and their applications: Collège de France Seminar. Volume III
H Brezis and J L Lions
71 Spectral theory and wave operators for the Schrödinger equation
A M Berthier
72 Approximation of Hilbert space operators I
D A Herrero
73 Vector valued Nevanlinna Theory
H J W Ziegler
74 Instability, nonexistence and weighted energy methods in fluid dynamics and related theories
B Straughan
75 Local bifurcation and symmetry
A Vanderbauwhede
76 Clifford analysis
F Brackx, R Delanghe and F Sommen
77 Nonlinear equivalence, reduction of PDEs to ODEs and fast convergent numerical methods
E E Rosinger
78 Free boundary problems, theory and applications. Volume I
A Fasano and M Primicerio
79 Free boundary problems, theory and applications. Volume II
A Fasano and M Primicerio
80 Symplectic geometry
A Crumeyrolle and J Grifone
81 An algorithmic analysis of a communication model with retransmission of flawed messages
D M Lucantoni
82 Geometric games and their applications
W H Ruckle
83 Additive groups of rings
S Feigelstock
84 Nonlinear partial differential equations and their applications: Collège de France Seminar. Volume IV
H Brezis and J L Lions
85 Multiplicative functionals on topological algebras
T Husain
86 Hamilton-Jacobi equations in Hilbert spaces
V Barbu and G Da Prato
87 Harmonic maps with symmetry, harmonic morphisms and deformations of metrics
P Baird
88 Similarity solutions of nonlinear partial differential equations
L Dresner
89 Contributions to nonlinear partial differential equations
C Bardos, A Damlamian, J I Díaz and J Hernández
90 Banach and Hilbert spaces of vector-valued functions
J Burbea and P Masani
91 Control and observation of neutral systems
D Salamon
92 Banach bundles, Banach modules and automorphisms of C*-algebras
M J Dupré and R M Gillette
93 Nonlinear partial differential equations and their applications: Collège de France Seminar. Volume V
H Brezis and J L Lions
94 Computer algebra in applied mathematics: an introduction to MACSYMA
R H Rand
95 Advances in nonlinear waves. Volume I
L Debnath
96 FC-groups
M J Tomkinson
97 Topics in relaxation and ellipsoidal methods
M Akgül
98 Analogue of the group algebra for topological semigroups
H Dzinotyiweyi
99 Stochastic functional differential equations
S E A Mohammed
100 Optimal control of variational inequalities
V Barbu
101 Partial differential equations and dynamical systems
W E Fitzgibbon III
102 Approximation of Hilbert space operators. Volume II
C Apostol, L A Fialkow, D A Herrero and D Voiculescu
103 Nondiscrete induction and iterative processes
V Ptak and F-A Potra
104 Analytic functions – growth aspects
O P Juneja and G P Kapoor
105 Theory of Tikhonov regularization for Fredholm equations of the first kind
C W Groetsch

106 Nonlinear partial differential equations and free boundaries
J I Díaz

107 Variational convergences for functions and operators
H Attouch

108 A layering method for viscous, incompressible L_p flows occupying R^n
A Douglis and E B Fabes

H A M Dzinotyiweyi

University of Zimbabwe

The analogue of the group algebra for topological semigroups

Pitman Advanced Publishing Program
BOSTON · LONDON · MELBOURNE

PITMAN PUBLISHING LIMITED
128 Long Acre, London WC2E 9AN

PITMAN PUBLISHING INC
1020 Plain Street, Marshfield, Massachusetts 02050

Associated Companies
Pitman Publishing Pty Ltd, Melbourne
Pitman Publishing New Zealand Ltd, Wellington
Copp Clark Pitman, Toronto

First published 1984

AMS Subject Classifications: (main) 43A20, 43A10, 4302, 4602
(subsidiary) 22A20

Library of Congress Cataloging in Publication Data

Dzinotyiweyi, H. A. M.
The analogue of the group algebra for topological semigroups.

Bibliography: p.
Includes indexes.
1. Topological semigroups. 2. Group algebras.
I. Title
QA387.D95 1984 512′.55 83-25003
ISBN 0-273-08610-3

British Library Cataloguing in Publication Data

Dzinotyiweyi, H. A. M.
The analogue of the group algebra for topological semigroups.—(Research notes in mathematics; 98)
1. Topological algebras
I. Title II. Series
512′.55 QA326

ISBN 0-273-08610-3

Reproduced and printed by photolithography
in Great Britain by Biddles Ltd, Guildford

Preface

A great deal of abstract harmonic analysis on a locally compact topological group G is pivoted at the object $L^1(G)$ - the so called group algebra of G. The purpose of this book is to present analogues of $L^1(G)$ for topological semigroups and, in particular, study those results whose success is largely due to the existence of such analogues on the underlying topological semigroup. Accordingly, this book is primarily addressed to research workers and post-graduate students in the field of abstract harmonic analysis on topological semigroups. Also, research workers in the fields of (theoretical) probability theory and functional analysis may find some useful information.

In Chapter 1 we collect together some preliminary results needed throughout this book. Most of the results are of an elementary nature. We also present the set of all bounded Radon measures on a topological semigroup as a convolution measure algebra.

In the absence of a Haar measure on a topological semigroup, one analogue of the group algebra is the set of all absolutely continuous bounded Radon measures, (i.e. bounded Radon measures whose translates are continuous with respect to compact sets). This analogue of the group algebra is studied in Chapter 2, mainly for topological semigroups admitting enough real-valued continuous functions to separate points. For many cases, it is shown that translates of absolutely continuous measures are weakly (and even norm) continuous. Another analogue of the group algebra for topological semigroups, namely the set of all bounded Radon measures which are equi-regular, is also studied in Chapter 2; in particular, we examine how far the latter differs from the set of all absolutely continuous measures. In general, the theme of Chapter 2 says that among the various analogues of the group algebra for topological semigroups, the set of all absolutely continuous measures seems to be the richest in structure.

In Chapter 3 various results on the topological structure of certain topological semigroups (which are generalizations of those) admitting "many" absolutely continuous measures are studied. For instance, for such semigroups, we show how far translates of neighbourhoods remain neighbourhoods. At the end of the Chapter, we characterize continuity of semigroup actions on normed linear spaces in terms of separable orbits, and the equivalence of weak and norm continuity of such actions is established.

In Chapter 4 various results on the space of weakly uniformly continuous functions WUC(S), the space of uniformly continuous functions UC(S), the space of weakly almost periodic functions WAP(S) and other spaces of functions on a topological semigroup S, admitting a non-zero absolutely continuous measure, are studied. Functions with separable orbits, which are also measurable with respect to every absolutely continuous measure, are shown to be "almost" uniformly continuous. For a large class of topological semigroups S we show that the existence of an invariant mean on A is equivalent to the existence of a topological invariant mean on B, for various spaces of functionals A and B on S where UC(S) is a subspace of $A \cap B$. Further, we show that the set of all invariant means on WUC(S) can be greater or equal to 2^c, where c denotes the cardinal of continuum. The difference of the spaces WAP(S) and WUC(S), for many topological semigroups S that include all noncompact locally compact topological groups, is shown to be large in a dramatic manner - in fact we show that the quotient space WUC(S)/WAP(S) may contain a linear isometric copy of ℓ^∞. We also examine the prospect of turning the Stone-Cech compactification of a topological semigroup S into a semitopological semigroup under an operation extending that of S. The Chapter ends with a study of the regularity of multiplication in measure algebras on semigroups and the size of the radical of the second dual of such algebras.

Chapter 5 is a natural continuation of Chapters 2 and 3. Here various characterizations of an absolutely continuous measure are given for the case of a topological semigroup studied in Chapter 3.

For a discrete semigroup S the analogues of the group algebra studied in this book, of course, coincide with the algebra $\ell^1(S)$. The latter algebra is still interesting, particularly as it often exhibits results full of algebraic content. We study some aspects of this algebra in Chapter 6. In particular we show that: there are many infinite semigroups S and weight functions ω such that the corresponding weighted algebra $\ell^1(S,\omega)$ has regular multiplication, the algebra $\ell^1(S)$ may be factorizable without an approximate identity, the existence of an approximate identity for $\ell^1(S)$ may be characterized in terms of an ordering type of property on the set of all idempotent elements of S. We also study the amenability of $\ell^1(S)$.

Strictly speaking, all our studies, from Chapter 1 to 6, fall under the field of Archimedean harmonic analysis, as all functions and measures involved pick their values in the Archimedean fields $\mathbb{R}$ or $\mathbb{C}$. In Chapter 7 we replace $\mathbb{R}$ and $\mathbb{C}$ by a non-Archimedean field K and present a Fourier theory for the resultant measure algebras.

Various open problems and detailed references are cited throughout the book. Some results employed, that do not strictly fall under the subject of abstract harmonic analysis, are assembled in Appendices A and B.

We have discussed and communicated with many colleagues on many matters presented in this book. In particular we would like to mention J.W. Baker, Ching Chou, Paul Milnes, A.L.T. Paterson, G.L.G. Sleijpen and A.C.M. van Rooij; whose company and correspondence we have always found to be very inspiring. Finally I wish to thank the referee for his invaluable comments and the Publishers for their cooperation.

University of Zimbabwe H.A.M.D.
November, 1983

Contents

1 Preliminaries

In this chapter we collect together some of the notation and terminology used throughout this book and include some standard and elementary results for ease of reference. We adopt the following nomenclature for references: The symbol [n] refers to reference number n in the bibliography list; notations of the form "Theorem n.m.k." refer to a theorem given in section m of chapter n under the label "m.k. THEOREM", and of the form "Theorem m.k." refer to a theorem given within the same chapter under the label "m.k. THEOREM". A similar convention is adopted for references to lemmas, propositions and other items.

1. SEMIGROUPS

1.1 DEFINITIONS. A nonempty set with an associative binary operation is called a *semigroup*. Let S be a semigroup. For all $A, B \subseteq S$ and $x \in S$ we write $AB := \{ab : a \in A \text{ and } b \in B\}$, $A^{-1}B := \{y \in S : ay \in B \text{ for some } a \in A\}$, $A^{-1}x := A^{-1}\{x\}$ and $x^{-1}B := \{x\}^{-1}B$. By symmetry our definitions for BA^{-1}, xA^{-1} and Bx^{-1} must be clear to the reader. We say S is *left* (or *right*) *cancellative* if $xy = xz$ (or $yx = zx$) for all $x,y,z \in S$ implies that $y = z$. S is said to be *cancellative* if it is both left and right cancellative. A point x in S is called a *left* (or *right*) *cancellation element* if it has the property that $xy = xz$ (or $yx = zx$) implies $y = z$, for all y,z in S. The *centre* Z(E) of a subset E of S is the set $Z(E) := \{s \in S : sx = xs \text{ for all } x \in E\}$. A non-empty subset H of S is called a *subsemigroup* if $HH = H$, *a left* (or *right*) *ideal* if $SH \subseteq H$ (or $HS \subseteq H$) and *an ideal* if it is both a left and right ideal.

Let e be an element of S. We say e is *an idempotent* if $e^2 = e$, a *zero* if $ex = xe = e$ for all x in S, a *left* (or *right*) *identity* if $ex = x$ (or $xe = x$) for all x in S and an *identity* (or *unity*) if it is both a left and right identity. The set of all idempotent elements in S is denoted by E_S.

By a *group* we mean a semigroup G with an identity element 1 such that corresponding to each x in G there is a unique y in G with the property $xy = yx = 1$. A non-empty subset of a semigroup S that is itself a group under the operation of S is called *a subgroup* of S.

1.2. Our reference for results on (algebraic) semigroups that are not proved or explicitly stated in this book is Clifford and Preston [20].

2. TOPOLOGICAL SEMIGROUPS

All topologies are assumed to be Hausdorff throughout this book.

2.1 DEFINITIONS. Let S be a semigroup endowed with a topology. Then S is called a *left* (or *right*) *semitopological semigroup* if the map $x \to xy$ (or $x \to yx$) of S into S is continuous, for all y in S. We say S is a *semitopological semigroup* if S is both a left and right semitopological semigroup. If the mapping $(x,y) \to xy$ of $S \times S$ into S is jointly continuous S is called a *topological semigroup*.

A group G is called a *topological group* if G is endowed with a topology with respect to which G is a topological semigroup and the inversion map $x \to x^{-1}$ of G into G is also continuous.

Let $A_1,\ldots,A_n$ be subsets of S. We define $A_1 \otimes A_2 := \{A_1A_2, A_1^{-1}A_2, A_1A_2^{-1}\}$, $A_1 \otimes A_2 \otimes A_3 := \cup\{A_1 \otimes B : B \in A_2 \otimes A_3\} \cup \cup\{B \otimes A_3 : B \in A_1 \otimes A_2\}$ and hence inductively similarly define $A_1 \otimes A_2 \otimes \ldots \otimes A_n$. A subset E of a topological semigroup S is said to be *relatively neo-compact* if E is contained in a union of sets in $A_1 \otimes \ldots \otimes A_n$ for some compact subsets $A_1,\ldots,A_n$ of S. We say S is *locally relatively neo-compact* if every point of S has a relatively neo-compact neighbourhood.

The reader should be in a position to easily solve the following exercises.

2.2 EXERCISES

(a) Every locally compact topological semigroup is locally relatively neo-compact.

(b) There exist locally relatively neo-compact topological semigroups that are not locally compact.

(c) Let S be a topological semigroup such that $C^{-1}D$ and DC^{-1} are compact whenever C and D are compact subsets of S. Then a subset E of S is relatively neo-compact if and only if E is relatively compact. (In particular, relatively neo-compact subsets of a topological group are precisely the relatively compact subsets.)

We say H is a *minimal ideal* of a semigroup S if H is an ideal contained in every other ideal of S. With this in mind we mention the following result without proof. (For a proof - see e.g. Hofmann and Mostert [62, pages 15 and 16].)

2.3 THEOREM. *Let* S *be a compact topological semigroup. Then there exists a minimal ideal* I *of* A. *Further,* $I \cap E_S$ *is non-empty and* eIe *is a topological subgroup of* S *with identity* e *whenever* $e \in I \cap E_S$.

One of the most important results in the theory of topological semigroups is the following.

2.4 THEOREM (Ellis [42, Theorem 2]). *Let* S *be a locally compact semi-topological semigroup. If* S *is algebraically a group, then* S *is a locally compact topological group.*

3. THE ALGEBRA OF BOUNDED RADON MEASURES M(S)

3.1 SOME GENERAL NOTATION ON FUNCTIONS. Recall that all topological spaces are assumed to be Hausdorff. Let X and Y be any topological spaces and S any topological semigroup.

We denote by $\mathcal{B}(X,Y)$ the set of all functions from X into Y. If E is a subset of X and f a function in $\mathcal{B}(X,Y)$, we take $f_{|E}$ to be the restriction of the function f to the set E. Let $\mathcal{C}_K(X,Y) := \{f \in \mathcal{B}(X,Y) : f_{|K}$ is continuous (on K) for all compact $K \subseteq X\}$ and $\mathcal{C}(X,Y) := \{f \in \mathcal{B}(X,Y) : f$ is continuous$\}$. The topological space X is said to be a *k-space* if $\mathcal{C}_K(X,Y) = \mathcal{C}(X,Y)$ for every topological space Y and a k_o-*space* if $\mathcal{C}_K(X,\mathbb{R}) = \mathcal{C}(X,\mathbb{R})$. (As illustrated in Chapter 4, k_o-spaces are not necessarily k-spaces.)

For all f in $\mathcal{B}(S,Y)$ and x in S we define the functions ${}_xf$ and f_x in $\mathcal{B}(S,Y)$ by

$${}_xf(y) := f(xy) \quad \text{and} \quad f_x(y) := f(yx), \quad \text{for all } y \text{ in } S.$$

Let $m(X) := \{f \in \mathcal{B}(X,\mathbb{C}) : f$ is bounded$\}$, $C(X) := m(X) \cap \mathcal{C}(X,\mathbb{C})$, $C_{oo}(X) := \{f \in C(X) : f$ vanishes outside a compact set$\}$ and $C_o(X) := \{f \in C(X) : f$ is arbitrarily small outside a compact set$\}$. We note that m(X) is a Banach space under the norm $\| \ \|_X$ given by $\|f\|_X := \sup\{|f(x)| : x \in X\}$. Whenever C(X) is treated as a normed space, it will be understood that $\| \ \|_X$ is the norm, unless otherwise stated. For any subset E of X we have the *characteristic function*, χ_E, of E given by

$$\chi_E(x) := \begin{cases} 1 & \text{if } x \in E \\ 0 & \text{if } x \in X \setminus E. \end{cases}$$

By the *support of a function* f in m(X) we mean the closure of the set $\{x \in X : f(x) \neq 0\}$.

The closure of a subset E in X is denoted by $\bar{E}$. If the real-valued functions in C(X) separate points of X, we say X is C-*distinguished*.

3.2 SOME GENERAL NOTATION ON MEASURES. Let X be any topological space and M(X) the space of all bounded complex-valued (inner and outer) regular Borel measures with the usual total variation norm $\| \ \|$. (Equivalently M(X) is the space of bounded Radon measures.) Corresponding to each μ in M(X) we have $|\mu|$ the measure arising from the total variation of μ and if E is a Borel subset of X then $\mu_{|E}$ is the measure given by $\mu_{|E}(B) := \mu(B \cap E)$ for all Borel $B \subseteq X$. If x is in X then $\bar{x}$ denotes the point mass at x. We also denote the completion of a measure μ by $\bar{\mu}$ (where the distinction from the notation for a point mass will lie in the context).

Let μ and ν be measures in M(X) and A a subset of M(X). By the *support* of μ we mean the set *supp*$(\mu) := \{x \in X : |\mu|(V) > 0$ whenever V is an open neighbourhood of x$\}$. We define the set function $|\mu|_*$ on subsets of X by $|\mu|_*(E) := \sup\{|\mu|(C) : C$ is a compact subset of E$\}$ and say μ is *concentrated* on a subset B of X if $|\mu|_*(B) = \|\mu\|$. Thus if μ is concentrated on a Borel subset B then $\mu_{|B} = \mu$. We say μ and ν are orthogonal, in short denoted $\mu \perp \nu$, if μ and ν are concentrated on disjoint Borel sets. Whenever μ is orthogonal to every measure in A we may denote this by $\mu \perp A$. A subset E of X is A-*measurable* if E is measurable with respect to every measure in A and A-*negligible* if E is negligible with respect to every measure in A. By the notation $\nu << \mu$ we mean that $\nu(E) = 0$ for every Borel set E with $\mu(E) = 0$. In particular we define $L(X,\mu) := \{\rho \in M(X) : \rho << \mu\}$. If μ is a positive measure then the space of all μ-integrable functions, $L^1(X,\mu)$, can be identified with $L(X,\mu)$, by the Radon-Nikodym Theorem. The set of all non-negative measures in A is denoted by A^+ and that of non-positive measures by A^-. We say A is solid if $\rho \in A$ whenever $\rho \in M(X)$ such that $\rho << |\mu|$ for some $\mu \in A$. Corresponding to A, we have a *density function* d_A given by $d_A(E) := \{x \in X$: for every neighbourhood V of x, we have $|\mu|_*(E \cap V) > 0$ for some $\mu \in A\}$, for all subsets E of X.

3.3 PROPOSITION. *Let X be a topological space, A a solid subspace of $M(X)$ and g a function in $m(X)$ which is A-measurable and such that* $\int g(x)d\nu(x) = 0$ *for all $\nu \in A$. Then g is A-almost everywhere equal to the zero function on X. In particular if $d_A(X) = X$ and $g \in C(X)$, then g is the zero function.*

Proof. We may assume that g is real-valued and define $g_1(x) := \max\{g(x),0\}$ and $g_2(x) := \max\{-g(x),0\}$, for all $x \in X$. Then $g = g_1 - g_2$, and $E_1 := \{x \in X : g_1(x) \neq 0\}$ and $E_2 := \{x \in X : g_2(x) \neq 0\}$ are disjoint A-measurable subsets. Since $|\mu|_{|E_1} \in A$ whenever $\mu \in A$, we have

$$\int g_1(x)d|\mu|(x) = \int g_1(x)d|\mu|_{|E_1}(x) = \int g(x)d|\mu|_{|E_1}(x) = 0 \qquad (\mu \in A).$$

So $g_1(x) = 0$ for $|\mu|$-almost every $x \in X$. Similarly $g_2(x) = 0$ and hence $g(x) = 0$ for $|\mu|$-almost every $x \in X$. Thus g is A-almost everywhere equal to zero. In particular, if $d_A(X) = X$ and $g \in C(X)$, then $g = 0$.

If M is a (complex) normed linear space we let M^* denote its continuous dual.

In this book, a topological semigroup S such that the real-valued functions in $C(S)$ separate points of S, is called a *C-distinguished topological semigroup*.

For any topological semigroup S, ν in $M(S)$ and f in $C(S)$, we define the functions νof and $fo\nu$ on S by

$$\nu of(x) := \int f(yx)d\nu(y) \quad \text{and} \quad fo\nu(x) := \int f(xy)d\nu(y) \qquad (x \in S).$$

We then have the following special case of Appendix B.9(i).

3.4 LEMMA. *If S is a topological semigroup, $\nu \in M(S)$ and $f \in C(S)$, then νof and $fo\nu$ are in $C(S)$.*

Blanket Assumption. *For the remainder of this section S denotes a C-cistinguished topological semigroup.*

We now commence an argument culminating in the establishment of a convolution operation on $M(S)$. For the details and notation on the strict topology we follow Appendix A.

3.5 LEMMA. *Let* μ_1 *and* μ_2 *be measures in* M(S) *and set*

$$L(f) := \iint f(xy)\,d\mu_1(x)\,d\mu_2(y)$$

for all f *in* C(S). *Then* L *is a strictly continuous linear functional on* C(S).

Proof. Indeed L is well defined as a linear mapping, by Lemma 3.4. We now show that L is strictly continuous. Without loss of generality we assume that μ_1 and μ_2 are positive measures and $\mu_1(S) = \mu_2(S) = 1$.

There are sequences of compact sets $\{K_{i,n}\}_{n=0}^{\infty}$ such that

$$\phi = K_{i,0} \subseteq K_{i,1} \subseteq \cdots \quad \text{and} \quad \mu_i(K_{i,n}) \geq 1-2^{-2n} \tag{1}$$

for i = 1,2. Let

$$\psi := \sum_{0\leq n,m<\infty} 2^{-(n+m)} \chi_{K_{1,n}K_{2,m}} .$$

Recalling definitions in Appendix A.1, we have that ψ is a hood.

Now, for all $x \in K_{1,n+1} \setminus K_{1,n}$ and $y \in K_{2,m+1} \setminus K_{2,m}$, we have

$$2^{-(m+n+2)} \leq \psi(xy). \tag{2}$$

Let ψ^{-1} be the (Borel) function defined on S by

$$\psi^{-1}(x) := \begin{cases} \dfrac{1}{\psi(x)} & \text{if } \psi(x) \neq 0 \\ 0 & \text{if } \psi(x) = 0. \end{cases}$$

We have

$$\begin{aligned} \iint \psi^{-1}(xy)\,d\mu_1(x)\,d\mu_2(y) &\leq \sum_{0\leq n,m<\infty} \int_{K_{2,n+1}\setminus K_{2,n}} \int_{K_{1,m+1}\setminus K_{1,m}} \psi^{-1}(xy)\,d\mu_1(x)\,d\mu_2(y) \\ &\leq \sum_{0\leq n,m<\infty} 2^{(n+m+2)} \mu_1(K_{1,m+1}\setminus K_{1,m})\,\mu_2(K_{2,n+1}\setminus K_{2,n}) \quad \text{by (2)} \\ &\leq \sum_{0\leq n,m<\infty} 2^{2-(n+m)} \quad \text{by (1)} \\ &= 16. \end{aligned}$$

Let $\varepsilon > 0$ be given and suppose $p_\psi(f) < \frac{1}{16}\varepsilon$ for some f in C(S). (For the definition of p_ψ see Appendix A.1.) Then

$$|f(xy)| < (\tfrac{1}{16}\varepsilon)\psi^{-1}(xy) \text{ for all } x \text{ in } \bigcup_{n=1}^{\infty} K_{1,n} \text{ and } y \text{ in } \bigcup_{n=1}^{\infty} K_{2,n}.$$

Consequently

$$|L(f)| \le \iint |f(xy)|d\mu_1(x)d\mu_2(y) \le (\tfrac{1}{16}\varepsilon)\iint \psi^{-1}(xy)d\mu_1(x)d\mu_2(x) \le \varepsilon.$$

Thus L is strictly continuous.

3.6. Appendix A.5 says that a linear functional L on C(S) is strictly continuous if and only if there exists a (unique) measure m in M(S) such that $L(f) = \int f\,dm$ for all f in C(S). In particular, recalling Lemma 3.5, it follows that corresponding to μ_1 and μ_2 in M(S) there is a unique measure $\mu_1 * \mu_2$ in M(S) such that

$$\iint f(xy)d\mu_1(x)d\mu_2(y) = \int f(x)d\mu_1{*}\mu_2(x)$$

for all f in C(S). With this notation we have the following result.

3.7 THEOREM. *Let* μ_1 *and* μ_2 *be any measures in* M(S). *Then for functions* f *in* $L^1(S, |\mu_1| * |\mu_2|)$, *we have that*

$$\int f(x)d\mu_1{*}\mu_2(x) = \iint f(zy)d\mu_1(z)d\mu_2(y) = \iint f(zy)d\mu_2(y)d\mu_1(z).$$

In particular for any Borel subset A *of* S *and* x *in* S, *we get*

$$\mu_1 * \mu_2(A) = \int \mu_1(Ay^{-1})d\mu_2(y) = \int \mu_2(z^{-1}A)d\mu_1(z),$$

$$\bar{x} * \mu_1(A) = \mu_1(x^{-1}A) \quad\text{and}\quad \mu_1 * \bar{x}(A) = \mu_1(Ax^{-1}).$$

In fact M(S) *is a Banach algebra with the convolution multiplication* $*$ *and the norm* $\|\ \|$.

Proof. The first assertion can be obtained by copying the proof of Hewitt and Ross [59, Theorem 19.10], on noting that the first paragraph of the proof can be adjusted accordingly by use of Appendix A.3(i), and the conclusion of the proof on [59, page 269] also follows in our case by use of Appendix A.3(ii). The part of our Theorem involving set A follows trivially by taking f := χ_A in the first part.

Now Appendix A.6(ii) says that M(S) is a Banach space under the norm $\| \ \|$. The convolution operation * is evidently associative, as defined. It is now easy to see that M(S) is a Banach algebra.

3.8 PROPOSITION. *Let* ν *and* μ *be positive measures in* M(S). *Then*

$$\operatorname{supp}(\nu * \mu) = \overline{\operatorname{supp}(\nu)\ \operatorname{supp}(\mu)} .$$

Proof. Let $z \in \operatorname{supp}(\nu*\mu)$. For any open neighbourhood V of z we thus have

$$0 < \nu*\mu(V) = \int \mu(x^{-1}V)\,d\nu(x).$$

So for some $x \in \operatorname{supp}(\nu)$ we have $\nu(x^{-1}V) > 0$ and so $x^{-1}V \cap \operatorname{supp}(\mu) \neq \phi$, which in turn implies that $V \cap x(\operatorname{supp}(\mu)) \neq \phi$ and $V \cap \operatorname{supp}(\nu)\ \operatorname{supp}(\mu) \neq \phi$. Thus $z \in \overline{\operatorname{supp}(\nu)\ \operatorname{supp}(\mu)}$.

Conversely let z be an element of $\overline{\operatorname{supp}(\nu)\ \operatorname{supp}(\mu)}$ and V any open neighbourhood of z. Then $V \cap \operatorname{supp}(\nu)\ \operatorname{supp}(\mu) \neq \phi$ and so there exist $x_1 \in \operatorname{supp}(\nu)$ and $x_2 \in \operatorname{supp}(\nu)$ with $x_1x_2 \in V$. Let V_1 and V_2 be open neighbourhoods of x_1 and x_2, respectively, such that $\overline{V_1V_2} \subset V$. Then

$$\nu*\mu(V) \geq \nu*\mu(\overline{V_1V_2}) \geq \int_{V_1} \mu(x^{-1}\overline{V_1V_2})\,d\nu(x) \geq \int_{V_1} \mu(V_2)\,d\nu(x) = \nu(V_1)\mu(V_2) > 0.$$

Hence $V \cap \operatorname{supp}(\nu*\mu) \neq \phi$ and so $z \in \operatorname{supp}(\nu*\mu)$.

3.9 DEFINITIONS. Let A be a subset of M(S). A is called a *convolution measure algebra* or an L-*subalgebra* if A is a norm-closed subalgebra of M(S) that is solid.

If A is an ideal of M(S) (i.e. $\nu*\mu$, $\mu*\nu \in A$ for all $\nu \in A$ and $\mu \in M(S)$) which is also an L-subalgebra, we call A an L-*ideal*. (For a more general theory of convolution measure algebras and L-ideals we refer the reader to e.g. J.L. Taylor [101].)

We say A is *left* (or *right*) *translation invariant* if $\bar{x}*\mu$ (or $\mu*\bar{x}$) is in A for all μ in A and x in S. Analogously a subset B of m(S) is *left* (or *right*) *translation invariant* if ${}_xf$ (or f_x) is in B for all f in B and x in S. A set is *translation invariant* if it is both left and right translation invariant.

The closure of the set $\cup\{\operatorname{supp}(\mu) : \mu \in A\}$ is called *the foundation of* A.

3.10 EXERCISES. Let A be a subset of M(S).

(i) The foundation of A coincides with the set $d_A(S)$.

(ii) If A is a (non-zero) subalgebra of M(S) then the foundation of A is a closed subsemigroup of S.

(iii) There is a C-distinguished topological group G such that $C_o(G) = \{0\}$.

(iv) There are locally compact topological semigroups such that neither $C_o(S)$ nor $C_{oo}(S)$ is translation invariant.

3.11 NOTATION. Let ν and μ be measures in M(S) and h a functional in M(S)*. We define functionals $\mu \circledcirc h$, $h \circledcirc \mu$ and $\nu \circledcirc h \circledcirc \mu$ in M(S)* by

$$\mu \circledcirc h(\eta) := h(\mu * \eta), \quad h \circledcirc \mu(\eta) := h(\eta * \mu) \quad \text{and} \quad \nu \circledcirc h \circledcirc \mu(\eta) := \nu \circledcirc h(\eta * \mu)$$

for all η in M(S). We define the functions $\mu o h$, $h o \mu$ and $\nu o h o \mu$ on S by

$$\mu o h(x) := \mu \circledcirc h(\bar{x}), \quad h o \mu(x) := h \circledcirc \mu(\bar{x}) \quad \text{and} \quad \nu o h o \mu(x) := \nu \circledcirc h \circledcirc \mu(\bar{x})$$

for all x in S.

If A is a subset of M(S)*, let $\mu \circledcirc A := \{\mu o h : h \in A\}$ and similarly define $A \circledcirc \mu$, $\mu \circledcirc A$ and $A \circledcirc \mu$.

3.12 NOTES. Proposition 3.8 is proved, for the case where S is locally compact, in the paper of J.S. Pym, "Idempotent measures on semigroups", *Pac. J. Math.* 12 (1962), 685-698. The notion of a foundation of a subset A of M(S) is taken from A.C. and J.W. Baker [5]. In fact following [5] we shall define foundation semigroups in Chapter 2. The other results in this section are taken from Dzinotyiweyi [30].

4. SOME CONTINUITY PROPERTIES OF RADON MEASURES

Throughout this section S *denotes any topological semigroup. We prove some results mainly needed in Chapter* 2.

4.1 LEMMA. *Let* μ *be a measure in* M(S) *and* E *a Borel subset of* S. *The map* $x \to |\mu|(x^{-1}E)$ *of* S *into* $\mathbb{R}$ *is*

(i) *lower semicontinuous when* E *is open,*

(ii) *upper semicontinuous when* E *is closed.*

Proof. Let E be open and (x_α) a net converging to a point x in S. Given $\varepsilon > 0$, we can choose a compact set $K \subseteq x^{-1}E$ such that $|\mu|(x^{-1}E \backslash K) < \varepsilon$. Now

$xK \subseteq x(x^{-1}E) \subseteq E$ and since E is open, the compactness of K implies the existence of an open neighbourhood V of x with $VK \subseteq E$. Hence $x_\alpha K \subseteq E$ eventually. Thus $K \subseteq x_\alpha^{-1}E$ eventually.

Consequently, eventually

$$|\mu|(x_\alpha^{-1}E) \geq |\mu|(K) = |\mu|(x^{-1}E) - |\mu|(x^{-1}E\backslash K)$$
$$> |\mu|(x^{-1}E) - \varepsilon.$$

This proves item (i). By complementation item (ii) follows.

4.2 PROPOSITION. *Let* μ *be a measure in* M(S), E *a compact subset of* S *and suppose* (x_α) *is a net converging to* x *in* S. *Then*

$$|\mu|(x_\alpha^{-1}E) \to |\mu|(x^{-1}E) \quad \textit{if and only if} \quad |\mu|(x_\alpha^{-1}E\Delta x^{-1}E) \to 0.$$

Proof. Since $\big||\mu|(x_\alpha^{-1}E) - |\mu|(x^{-1}E)\big| \leq |\mu|(x_\alpha^{-1}E\Delta x^{-1}E)$, it is sufficient to show that $|\mu|(x_\alpha^{-1}E\Delta x^{-1}E) \to 0$ if $|\mu|(x_\alpha^{-1}E) \to |\mu|(x^{-1}E)$. To this end, given $\varepsilon > 0$, we have that eventually

$$\big||\mu|(x_\alpha^{-1}E) - |\mu|(x^{-1}E)\big| < \tfrac{1}{2}\varepsilon$$

and, by Lemma 4.1(ii),

$$|\mu|(x_\alpha^{-1}E\backslash x^{-1}E) = |\mu|_{|S\backslash x^{-1}E}(x_\alpha^{-1}E) < \tfrac{1}{4}\varepsilon.$$

Hence

$$|\mu|(x_\alpha^{-1}E\Delta x^{-1}E) = |\mu|(x_\alpha^{-1}E\backslash x^{-1}E) + |\mu|(x^{-1}E\backslash x_\alpha^{-1}E)$$
$$= 2|\mu|(x_\alpha^{-1}E\backslash x^{-1}E) + |\mu|(x^{-1}E) - |\mu|(x_\alpha^{-1}E)$$
$$\leq 2|\mu|(x_\alpha^{-1}E\backslash x^{-1}E) + \big||\mu|(x^{-1}E) - |\mu|(x_\alpha^{-1}E)\big|$$
$$< \varepsilon \quad \text{eventually.}$$

The preceding items are prove for the case where S is locally compact in [5] and [92], respectively. The actual setting given above is taken from [30].

2 Analogues of the group algebra for semigroups

We introduce and prove elementary properties of certain measure algebras on topological semigroups. These measure algebras coincide with the usual group algebra $L^1(G)$ in the case of a locally compact topological group G.

1. ABSOLUTELY CONTINUOUS MEASURES

Let S be a topological semigroup and recall that M(S) is the space of all bounded Radon measures on S. In this section we shall be concerned with the following subspaces of M(S) :

$$M_a^\ell(S) := \{\mu \in M(S) : \text{the map } x \to |\mu|(x^{-1}C) \text{ of } S \text{ into } \mathbb{R} \text{ is continuous for all compact } C \subset S\}$$

$$M_a^r(S) := \{\mu \in M(S) : \text{the map } x \to |\mu|(Cx^{-1}) \text{ of } S \text{ into } \mathbb{R} \text{ is continuous for all compact } C \subset S\}$$

$$M_a(S) := M_a^\ell(S) \cap M_a^r(S).$$

In particular we say a measure is *absolutely continuous* on S if it belongs to the set $M_a(S)$.

1.1 THEOREM. *If* S *is a* C-*distinguished topological semigroup, then* $M_a^\ell(S)$ *is an* L-*subalgebra and also a left ideal of* M(S).

Proof. Let $\nu \in M(S)$ and $\mu \in M_a^\ell(S)$ with $\nu << |\mu|$ be given, and suppose that (x_α) is a net converging to x in S and C is a compact subset of S. Then lemma 1.4.2 says that $|\mu|(x_\alpha^{-1}C\Delta x^{-1}C) \to 0$ and so $|\nu|(x_\alpha^{-1}C\Delta x^{-1}C) \to 0$, by the Radon-Nikodym theorem. Hence $\nu \in M_a^\ell(S)$ also by lemma 1.4.2.

Since $M_a^\ell(S)$ is solid, to show that $M_a^\ell(S)$ is norm closed it is sufficient to prove that if $\{\mu_n\}$ is a sequence of positive measures in $M_a^\ell(S)$ converging to $\mu \in M(S)$, then $\mu \in M_a^\ell(S)$. Noting that μ is then a positive measure, that $\mu \in M_a^\ell(S)$ follows easily from our definition of $M_a^\ell(S)$.

So to complete the proof it remains to show that $M_a^\ell(S)$ is a left ideal of M(S). Let $\nu \in M(S)$ and $\mu \in M_a^\ell(S)$ be fixed, $\varepsilon > 0$ be given, C be a compact subset of S and suppose that (x_α) is a net converging to x in S. We can choose a compact set $K \subset S$ such that

$$\|\nu - \nu_{|K}\| < \frac{\varepsilon}{2(1+\|\mu\|)} .$$

Since $\mu \in M_a^\ell(S)$, a standard argument shows that, eventually

$$||\mu|((x_\alpha y)^{-1}C) - |\mu|((xy)^{-1}C)| < \frac{\varepsilon}{2(1+|\nu|(K))} \quad (y \in K).$$

Thus eventually we get

$$\begin{aligned} ||\nu|*|\mu|(x_\alpha^{-1}C) - |\nu|*|\mu|(x^{-1}C)| &= |\int(\bar{y}*|\mu|(x_\alpha^{-1}C) - \bar{y}*|\mu|(x^{-1}C))d|\nu|(y)| \\ &\leq \int||\mu|((x_\alpha y)^{-1}C) - |\mu|((xy)^{-1}C)|d|\nu|(y) \\ &\leq \int_K ||\mu|((x_\alpha y)^{-1}C) - |\mu|((xy)^{-1}C)|d|\nu|(y) + \tfrac{1}{2}\varepsilon \\ &< \tfrac{1}{2}\varepsilon + \tfrac{1}{2}\varepsilon = \varepsilon. \end{aligned}$$

So $|\nu|*|\mu| \in M_a^\ell(S)$ and, since $\nu*\mu << |\nu|*|\mu|$, we have $\nu*\mu$ in $M_a^\ell(S)$. Thus $M_a^\ell(S)$ is a left ideal.

<u>Remark</u>. Note that we do not use the fact that S is C-distinguished in the first paragraph of our proof. Thus for every topological semigroup S if $\nu << |\mu|$ and $\mu \in M_a^\ell(S)$, then $\nu \in M_a^\ell(S)$ (i.e. $M_a^\ell(S)$ is solid).

We needed S to be C-distinguished in order to define the convolution operation on M(S). For such topological semigroups S we will show that $M_a(S)$ is always an L-ideal. Towards this end we first prove some lemmas.

In the following sequence of lemmas let S *be a* C-*distinguished topological semigroup.*

1.2 LEMMA. *If* $\mu \in M_a^\ell(S)$ *and* E *is a Borel subset of* S, *then the map* $x \to \mu(x^{-1}E)$ *of* S *into* $\mathbb{C}$ *is continuous.*

Proof. By the Hahn decomposition theorem and since $M_a^\ell(S)$ is solid it is sufficient to prove the lemma for the case where μ is a positive measure.

So suppose μ is positive, (x_α) is a net converging to x in S and let $\varepsilon > 0$ be given. We can find a compact set

$C \subseteq x^{-1}E$ with $\mu(C) > \mu(x^{-1}E) - \varepsilon$.

Now $xC \subseteq x(x^{-1}E) \subseteq E$ and hence $x_\alpha^{-1}(xC) \subseteq x_\alpha^{-1}E$. Consequently

$$\underline{\lim}_\alpha \ \mu(x_\alpha^{-1}E) \geq \lim_\alpha \mu(x_\alpha^{-1}(xC)) = \mu(x^{-1}(xC)) \geq \mu(x^{-1}E) - \varepsilon. \tag{1}$$

Hence if E were closed, (i) and Lemma 1.4.1(ii) would give the result. Then by complementation, our lemma also follows for open E.

But, by the outer regularity of $\bar{x}*\mu$, we can choose an open set $U \supseteq E$ with

$$\bar{x}*\mu(U) < \bar{x}*\mu(E) + \varepsilon.$$

So with (1) in mind we get

$$\begin{aligned}\bar{x}*\mu(E) - \varepsilon &\leq \underline{\lim}_\alpha \ \bar{x}_\alpha*\mu(E) \leq \overline{\lim}_\alpha \ \bar{x}_\alpha*\mu(E) \\ &\leq \lim_\alpha \bar{x}_\alpha*\mu(U) = \bar{x}*\mu(U) < \bar{x}*\mu(E) + \varepsilon.\end{aligned}$$

This completes our proof.

1.3 NOTATION. For any subset E of S and measure μ in M(S) we define the maps $\ell_{\mu,E}$ and $r_{\mu,E}(x)$ of E into M(S) by

$$\ell_{\mu,E}(x) := \bar{x}*\mu \quad \text{and} \quad r_{\mu,E}(x) := \mu*\bar{x} \quad \text{for all } x \in E,$$

and *whenever we talk of the continuity of the maps* $\ell_{\mu,E}$ *and* $r_{\mu,E}$ *it will be understood that the set* E *has the relative topology of* S. We also define $\ell_\mu := \ell_{\mu,S}$ and $r_\mu := r_{\mu,S}$.

1.4 LEMMA. *Let* μ *be a measure in* $M_a^\ell(S)$ *and* K *a compact subset of* S. *Then* $\ell_{\mu,K}$ *is a weakly continuous mapping of* K *into* M(S).

Proof. Since $M_a^\ell(S)$ is solid and the space $A_w := \{\nu \in M(S) : \ell_{\nu,K}$ is weakly continuous$\}$ is clearly norm closed, it is sufficient to prove the lemma for the case where $H := \mathrm{supp}(\mu)$ is compact.

So assuming H to be compact we evidently have

$$\mathrm{supp}(\bar{x}*\mu) \subseteq KH \quad \text{for all} \quad x \in K.$$

Clearly Lemma 1.2 implies that $A := \{\bar{x}*\mu : x \in K\}$ satisfies condition (ii) of

Grothendieck's Theorem B.4 (in Appendix) and hence A is a relatively weakly compact subset of M(KH).

Let (x_α) be a net in K converging to some point $x \in K$. Then $(\bar{x}_\alpha * \mu)$ has a weak cluster point, say, ρ in M(KH). So there is a subnet (y_γ) of (x_α) such that $(\bar{y}_\gamma * \mu)$ converges weakly to ρ in M(KH). In particular if E is any Borel subset of S; then $\bar{y}_\gamma * \mu(E)$ converges to $\rho(E)$. But $\bar{x}_\alpha * \mu(E)$ converges to $\bar{x} * \mu(E)$, by Lemma 1.2. Hence $\rho(E) = \bar{x} * \mu(E)$ and $(\bar{x}_\alpha * \mu)$ has one weak cluster point $\bar{x} * \mu$. Thus $\ell_{\mu,K}$ is a weakly continuous mapping of K into M(KH).

Observing that if $h \in M(S)^*$, the restriction of h to M(KH) belongs to $M(KH)^*$, our lemma follows.

1.5 LEMMA. *Let* μ *be any measure in* $M_a^\ell(S)$ *and* $h \in M(S)^*$. *Then the map* $x \to h(\bar{x} * \mu)$ *defines a function on* S *that is measurable with respect to every measure in* M(S).

Proof. Let ν be any positive measure in M(S) and choose and increasing sequence of compact sets $\{K_n\}$ such that $\|\nu\| = \nu(\bigcup_{n=1}^{\infty} K_n)$. Now by Lemma 1.4 the map $x \to \chi_{K_n}(x)\, h(\bar{x} * \mu)$ defines a ν-measurable function on S. Consequently taking $E := \bigcup_{n=1}^{\infty} K_n$ we have the map $x \to \chi_E(x) h(\bar{x} * \mu)$ ν-measurable on S. Observing that ν vanishes outside E, we have that $x \to h(\bar{x} * \mu)$ is a ν-measurable function. Hence our lemma follows.

A measure $\mu \in M(S)$ is said to be *left* (*right*) *equi-absolutely continuous* if $\bar{x} * |\mu| << |\nu| * |\mu|$ ($|\mu| * \bar{x} << |\mu| * |\nu|$, respectively), for all $x \in supp(\nu)$ and $\nu \in M(S)$. We say μ is *equi-absolutely continuous* if μ is both left and right equi-absolutely continuous. We shall study such measures later in Chapter 5, Section 2. For the moment we take note of Lemma 1.6(i).

1.6 LEMMA. *Let* $\nu \in M(S)$, $\mu \in M_a^\ell(S)$ *and* $h \in M(S)^*$. *Then*

(i) $\bar{x} * |\mu| << |\nu| * |\mu|$ *for all* $x \in supp(\nu)$. *(Thus* μ *is left equi-absolutely continuous.)*

(ii) $h(\nu * \mu) = \int h(\bar{x} * \mu) d\nu(x)$.

Proof. (i) Suppose $|\nu|*|\mu|(E) = 0$ for some Borel subset E of S. Then

$$\int \bar{x}*|\mu|(E)\,d|\nu|(x) = 0$$

and so $\bar{x}*|\mu|(E) = 0$ for ν-almost every $x \in \text{supp}(\nu)$. Recalling Lemma 1.2 we have $\bar{x}*|\mu|(E) = 0$ for all $x \in \text{supp}(\nu)$.

(ii) Since h is a continuous linear functional on $L(S,|\nu|*|\mu|)$, the Radon-Nikodym Theorem implies the existence of a bounded Borel measurable function f on S such that

$$h(\rho) = \int f(x)\,d\rho(x) \quad \text{for all} \quad \rho \in L(S,|\nu|*|\mu|).$$

Now from (i) it is trivial to note that $\bar{x}*\mu$, $\nu*\mu \in L(S,|\nu|*|\mu|)$ for all $x \in \text{supp}(\nu)$.

So

$$\begin{aligned} h(\nu*\mu) &= \int f(x)\,d\nu*\mu(x) = \iint f(xy)\,d\nu(x)\,d\mu(y) \\ &= \int\left\{\int f(xy)\,d\mu(y)\right\}d\nu(x) \quad \text{by Fubini's Theorem} \\ &= \int \{f(z)\,d(\bar{x}*\mu)(y)\}\,d\nu(x) \\ &= \int h(\bar{x}*\mu)\,d\nu(x). \end{aligned}$$

1.7 THEOREM. *If* S *is a* C-*distinguished topological semigroup then* $M_a(S)$ *is an* L-*ideal of* M(S).

Proof. Lemma 1.2 implies that $|\mu|*\bar{x} \in M_a^{\ell}(S)$ for all $\mu \in M_a^{\ell}(S)$ and $x \in S$. Since $\mu*\bar{x} << |\mu|*\bar{x}$ and $M_a^{\ell}(S)$ is an L-subalgebra we have $\mu*\bar{x} \in M_a^{\ell}(S)$. Since $M_a^{\ell}(S)$ is a left ideal (recall Theorem 1.1) we thus have $\bar{x}*\mu$, $\mu*\bar{x} \in M_a^{\ell}(S)$ and hence $M_a^{\ell}(S)$ is translation invariant. Similarly $M_a^{r}(S)$ is translation invariant and hence $M_a(S)$ is translation invariant.

It is trivial to note that $M_a(S)$ is already an L-subalgebra, by Theorem 1.1. To show that $M_a(S)$ is an ideal let $\nu \in M(S)$ and $\mu \in M_a(S)$ be fixed and suppose on the contrary $\nu*\mu \notin M_a(S)$. Then the Hahn-Banach Theorem implies the existence of a functional $h \in M(S)^*$ vanishing on $M_a(S)$ but with $h(\nu*\mu) = 1$. By Lemma 1.6(ii) this implies that

$$1 = h(\nu*\mu) = \int h(\bar{x}*\mu)\,d\nu(x) = 0 \quad (\text{since } \bar{x}*\mu \in M_a(S))$$

- conflict! By this conflict we conclude that $\nu*\mu \in M_a(S)$ and similarly $\mu*\nu \in M_a(S)$. This completes our proof.

Next we show that the existence of many (left or right) absolutely continuous measures on a topological semigroup implies that the semigroup is C-distinguished, for many cases.

First we recall the following notation: *an element* e *of a semigroup* S *is called a right cancellation element if* ye = ze *implies that* y = z *for all* y, z ∈ S. In particular if S has an identity element 1, then 1 is a right cancellation element.

1.8 THEOREM. *Let* S *be any topological semigroup containing a right cancellation element* e. *Then if* S *coincides with the foundation of* $M_a^{\ell}(S)$ *we have that* S *is C-distinguished.*

Proof. Let x and y be any distinct points of S. Then ye ≠ ze and so, since S is Hausdorff, we can find a neighbourhood V of e such that $yV \cap zV = \phi$. By the remark following Theorem 1.1, we can choose a positive measure $\nu \in M_a^{\ell}(S)$ with K := supp(ν) compact, $K \subseteq V$ and ν(K) = 1. Now the function $f : x \to \nu(x^{-1}(yK))$ is continuous and bounded on S, and

$$f(y) = \nu(y^{-1}(yK)) = \nu(K) = 1$$

while $f(z) = \nu(z^{-1}(yK)) < \nu(z^{-1}(yV)) = \nu(z^{-1}(yV) \cap V) = 0.$

So S is C-distinguished.

For the remainder of this section, apart from item 1.15, we examine the nature of the topology of those topological semigroups admitting absolutely continuous measures. In particular we recall that for a topological group G, existence of an absolutely continuous measure on G implies that G is locally compact - see e.g. Gowrisankaran [48], and we also achieve this result in item 1.10.

1.9 LEMMA. *Let* S *be any topological semigroup and* $\nu \in M_a^{\ell}(S)$ *a positive measure with* D := supp(ν) *compact. Then* $(xD)D^{-1}$ *is a neighbourhood of* x *for all* x ∈ S.

Proof. By definition of $M_a^{\ell}(S)$ we have that $O(x) := \{s \in S : \nu(s^{-1}(xD) > 0\}$ is an open neighbourhood of x, for all $x \in S$. Now

$$O(x) \subseteq \{s \in S : s^{-1}(xD) \cap D \neq \phi\} = (xD)D^{-I}.$$

Recalling the definition of relative neo-compactness from item 1.2.1 we have

1.10 COROLLARY. *Let* S *be a topological semigroup with* $M_a^{\ell}(S)$ *non-zero. Then* S *is locally relatively neo-compact. In particular if* CD^{-1} *is compact whenever* C *and* D *are compact subsets of* S, *then* S *is locally compact.*

1.11. For any topological semigroup S, recalling the notation of item 1.3.1 we define $S_r := \{x \in S :$ if $f \in C_K(S,Y)$ then $f_x \in C(S,Y)$, for every Hausdorff topological space $Y\}$. Note that if S has an identity element 1 then $1 \in S_r$ if and only if S is a k-space. More generally we may say that the "larger" S_r is the "closer" S is to being a k-space. If we accept this then the next three items, together, show that existence of "many" absolutely continuous measures on S implies that S is "almost" a k-space.

For the remainder of this section S *denotes a* C-*distinguished topological semigroup unless otherwise stated.*

1.12 THEOREM. *If* H *denotes the foundation of* $M_a^{\ell}(S)$ *then* $H^2 \subseteq \bar{S}_r$.

Proof. Let $z_1, z_2 \in H$ and U be any neighbourhood of $z := z_1z_2$. We can choose open neighbourhoods V_1,V_2 of z_1,z_2 (respectively) such that $V_1V_2 \subseteq U$. Since $M_a^{\ell}(S)$ is an L-subalgebra (by Theorem 1.1) we can find positive measures ν_1 and ν_2 in $M_a^{\ell}(S)$ with $C_1 := \mathrm{supp}(\nu_1)$ and $C_2 := \mathrm{supp}(\nu_2)$ compact and such that $C_1 \subseteq V_1$ and $C_2 \subseteq V_2$. Now observing that $0 < \bar{x}*\nu_2(xC_2)$ for all $x \in S$ and that $\bar{x}*\nu_2 << \nu_1*\nu_2$, by Lemma 1.2, we obtain

$$0 < \int \nu_1*\nu_2(xC_2)d\nu_1(x) = \iint \chi_{xC_2}(y)d\nu_1(x)d\nu_1*\nu_2(y) \text{ by Fubini's theorem}$$

$$= \iint \chi_{yC_2^{-1}}(x)d\nu_1(x)d\nu_1*\nu_2(y)$$

$$= \int \nu_1(yC_2^{-1})d\nu_1*\nu_2(y).$$

So $\nu_1(y_oC_2^{-1}) > 0$ for some $y_o \in \mathrm{supp}(\nu_1*\nu_2) \subseteq C_1C_2 \subseteq V_1V_2 \subseteq U$.

Now, as in the proof of Lemma 1.9, since $\nu_{1|y_oC_2^{-1}} \in M_a^{\ell}(S)$ we have that

$$O(s) := \{x \in S : \nu_{1|y_oC_2^{-1}}(x^{-1}(sC_1)) > 0\}$$

is a neighbourhood of s and clearly

$$O(s) \subseteq \{x \in S : x^{-1}(sC_1) \cap y_oC_2^{-1} \neq \phi\}$$

$$= (sC_1)(y_oC_2^{-1})^{-1} \subseteq (sC_1C_2)y_o^{-1}. \quad (s \in S)$$

Now suppose that $f \in C_K(S,Y)$ for some Hausdorff topological space Y, and let (s_α) be a net converging to s in S. Then since $(sC_1C_2)y_o^{-1}$ is a neighbourhood of s we have that eventually $(s_\alpha y_o)$ is a net converging to sy_o in the compact set sC_1C_2. Hence $f(s_\alpha y_o) \to f(sy_o)$ or $f_{y_o}(s_\alpha) \to f_{y_o}(s)$. Hence $f_{y_o} \in C(S,Y)$ and so $y_o \in S_r \cap U$. It follows that $z \in \bar{S}_r$ and so $H^2 \subseteq \bar{S}_r$.

From Lemma 1.4 and Theorem 1.12 we have the following result.

1.13 COROLLARY. *If* H *is the foundation of* $M_a^{\ell}(S)$ *then* H^2 *is contained in the closure of the set* $\{x \in S : \ell_{\bar{x}*\mu}$ *is a weakly continuous mapping of* S *into* M(S), *for all* $\mu \in M_a(S)\}$.

1.14 THEOREM. *We have that*

(a) S_r *is a left ideal of* S.

(b) *If* S *is the foundation of* $M_a(S)$ *and* $\mu \in M_a(S)$, *then* $\mu*\bar{x}$ *is concentrated on* S_r, *for all* $x \in S^2$. *In particular, if in addition* S *has an identity element, then* $M_a(S)$ *is concentrated on* S_r.

Proof. That S_r is a left ideal follows trivially from our definitions.

We now prove item (b). Let $\mu \in M_a(S)$ be fixed, D := supp(μ) and K be any compact subset of $S \setminus S_r$. Then, for all $x \in S_r$ we have $Dx \subseteq S_r$, since S_r is a left ideal and hence

$$|\mu|*\bar{x}(K) = 0.$$

Now $S^2 \subseteq \bar{S}_r$, by Theorem 1.12. So, by the definition of $M_a(S)$, we must have $|\mu|*\bar{x}(K) = 0$ for all $x \in S^2$. Since $\mu*\bar{x} << |\mu|*\bar{x}$, the proof for the first part is complete. The second part follows trivially from the first part.

Next we mention some specific situations where S admits absolutely continuous measures, and also cases (e.g. Examples 1.15(f)) justifying our definitions.

1.15 EXAMPLES

(a) If S is discrete then $M_a(S) = M(S) = \ell^1(S)$ - the algebra introduced by Hewitt and Zuckerman [60]. We shall study some aspects of this algebra in Chapter 6.

(b) *If* S *admits a positive regular Borel measure* m *such that* $m(x^{-1}E) \leq m(E)$ *for all* $x \in S$ *and Borel* $E \subseteq S$, *then* $L(S,m) \subseteq M_a^\ell(S)$.

Proof. Let $\nu \in L(S,m)$ and $\varepsilon > 0$ be given. We can find a $\delta > 0$ such that for every Borel $E \subseteq S$ with $m(E) < \delta$ we have $|\nu|(E) < \varepsilon$. Let (x_α) be a net converging to x in S and K be a compact subset of S. We can find an open neighbourhood O of K such that $m(O\setminus K) < \delta$. Then $m(s^{-1}(O\setminus K)) < \delta$ and hence $|\nu|(s^{-1}(O\setminus K)) < \varepsilon$ for all $s \in S$. So (eventually)

$$\begin{aligned}\bar{x}_\alpha*|\nu|(K) &= \bar{x}_\alpha*|\nu|(O) - \bar{x}_\alpha*|\nu|(O\setminus K)\\ &\geq \bar{x}_\alpha*|\nu|(O) - \varepsilon\\ &> \bar{x}*|\nu|(O) - 2\varepsilon \qquad \text{by Lemma 1.4.1(i)}\\ &> \bar{x}*|\nu|(K) - 2\varepsilon .\end{aligned}$$

Recalling Lemma 1.4.1(ii), we conclude that $L(S,m) \subseteq M_a^\ell(S)$.

(c) *Let* S *be a topological group. Then either* $M_a^\ell(S)$ *is zero or* S *is locally compact and* $M_a^\ell(S) = M_a^r(S) = L(S,m)$ *where* m *is a left Haar measure on* S. (cf. [48] and [59]).

Proof. That if $M_a^\ell(S)$ is non-zero then S is locally compact follows from Corollary 1.10. Let m_1 be a right Haar measure on G. Item (b) above implies that $L(S,m) \subseteq M_a^\ell(S)$ and $L(S,m_1) \subseteq M_a^r(S)$.

Let $\mu \in M_a^\ell(S)$ and V be a compact neighbourhood of the identity of S. Then clearly $\mu << m_{|V}*|\mu|$. For if K is any compact set with $m_{|V}*|\mu|(K) = 0$, then $0 = \int \bar{x}*|\mu|(K)dm_{|V}(x)$; so $\bar{x}*|\mu|(K) = 0$, for all $x \in V$, and (in particular) $|\mu|(K) = 0$. By the right invariance of m_1 we also have $m_{1\,V}*|\mu| << m_1$. Hence $\mu << m_1$ and $\mu \in L(S,m_1)$. But it is well known (see e.g. Hewitt and Ross [59]) that $L(S,m) = L(S,m_1)$. So $\mu \in L(S,m)$ and

$M_a^\ell(S) = L(S,m)$. The rest of the proof should now be clear.

(d) Let P be the set of all rational numbers in $[0,\infty)$ and take

$$S := \{(x,0) : x \in P\} \cup \{(x,y) : x \in [0,\infty),\ y \in (0,\infty)\}$$

the additive subsemigroup of $\mathbb{R}^2$ with the usual topology. Thus S is a cancellative commutative C-distinguished topological semigroup that is not locally compact. S is also a k-space and one easily notes that S is the foundation of $M_a(S)$. Thus foundations of absolutely continuous measures (unlike the group case) may not be locally compact even for cases where the underlying subsemigroup is a topological subsemigroup of a locally compact topological group.

(e) Let X be the direct product of uncountably many copies of $\mathbb{R}$ with the product topology. Then X is a C-distinguished (in fact a Tychonoff) space but not a k-space. Fix $z \in X$ and take S to be X with the multiplication given by $xy := z$ for all $x,y \in X$. Then $M_a(S) = M(S)$. Thus for the foundation of absolutely continuous measures to be a k-space one needs to impose some conditions on the underlying semigroup.

(f) Let H be the semigroup $[0,1]$ with maximum operation and the usual topology. Take S to be the (commutative and compact) topological semigroup $H \cup \{e\}$ where e is an adjoined isolated identity element. Consider the measure $\mu := \bar{0} - \bar{e}$ and note that

$$\bar{x}*\mu = \bar{x}*\bar{0} - \bar{x}*\bar{e} = \bar{x} - \bar{x} = 0 \quad \text{for all } x \in H$$

and so the map $x \to \bar{x}*\mu$ of S into M(S) is norm continuous and for all compact $K \subseteq S$ we have $x \to \bar{x}*\mu(K)$ continuous. Further $\bar{0} << |\mu|$, $\bar{0} \notin M_a^\ell(S)$ and $x \to \bar{x}*\bar{0}$ is not a norm continuous map [5].

2. WEAK AND NORM CONTINUOUS TRANSLATIONS OF MEASURES

Throughout this section S *stands for a* C-*distinguished topological semigroup.*

We examine the weak and norm continuity of functions ℓ_μ and r_μ for any absolutely continuous measure μ and shade some light on the spaces

$$M_n^\ell(S) := \{\mu \in M(S) : \ell_\mu \text{ is norm continuous}\}$$

$$M_n^r(S) := \{\mu \in M(S) : r_\mu \text{ is norm continuous}\}$$

$$M_n(S) := M_n^\ell(S) \cap M_n^r(S)$$

first introduced by Hart [58] for locally compact S.

For ease of reference we mention the following two lemmas. The proof of our first lemma is contained in that of Lemma 1.4.

2.1 LEMMA. *If* K *is a compact subset of* S *and* $\mu \in M(S)$, *the following items are equivalent:*

(i) *the map* $\ell_{\mu,K}$ *is weakly continuous;*
(ii) *the set* $\{\bar{x}*\mu : x \in K\}$ *is (relatively) weakly compact.*

In a like manner the reader should be able to furnish a proof for the next lemma.

2.2 LEMMA. *If* K *is a compact subset of* S *and* $\mu \in M(S)$, *then the following items are equivalent:*

(i) *the map* $\ell_{\mu,K}$ *is norm continuous;*
(ii) *the set* $\{\bar{x}*\mu : x \in K\}$ *is (relatively) norm compact.*

For locally compact S the following result is due to Sleijpen [92].

2.3 THEOREM. *Let* ν *and* μ *be any measures in* $M_a^\ell(S)$ *and* K *a compact subset of* S. *Then* $\ell_{\nu*\mu,K}$ *is norm continuous.*

Proof. Since $M_a^\ell(S)$ is solid and one can easily note that $M_n^\ell(S)$ is closed, it is sufficient to give a formal proof for the case where $\mathrm{supp}(\nu)$ and $\mathrm{supp}(\mu)$ are compact.

So suppose $H := \mathrm{supp}(\nu)$ and $L := \mathrm{supp}(\mu)$ are compact. We note that the measure $\bar{x}*|\mu|$ is concentrated on the compact set KHL for all $x \in KH$ and $\bar{y}*\nu$ is concentrated on the compact set KH for all $y \in K$.

Now if A is any bounded subset of M(KH), then Lemma 1.4 easily enables one to show that $\{|\eta|*|\mu| : \eta \in A\}$ satisfies condition (ii) of Grothendieck's Theorem B.4 (in Appendix) and hence, by the same Theorem, we have that $\{\eta*\mu : \eta \in A\}$ is a relatively weakly compact subset of M(KHL). Consequently the mapping P_μ of M(KH) into M(KHL) given by

$$P_\mu(\rho) := \rho*\mu \quad \text{for all} \quad \rho \in M(KH),$$

is a weakly compact operator.
Since $\{\bar{x}*\nu : x \in K\}$ is a weakly compact subset of M(KH), (recall Lemma 1.4), the Dunford-Pettis Property (see Appendis B.6,) says that

$$P_\mu(\{\bar{x}*\nu : x \in K\}) = \{\bar{x}*\nu*\mu : x \in K\}$$

is a norm compact subset of M(KHL). Hence by Lemma 2.2, $\ell_{\nu*\mu,K}$ is a norm continuous mapping of K into M(KHL). So noting that $\bar{x}*\nu*\mu$ is concentrated on KHL for all $x \in K$, we are done.

Recalling Lemma 1.4, we have the following result.

2.4 COROLLARY. *Let* ν *and* μ *be measures in* $M_a^\ell(S)$ *and suppose that* S *is also a* k-*space. Then*

(i) ℓ_ν *is weakly continuous;*
(ii) $\ell_{\nu*\mu}$ *is norm continuous.*

Thus there seems to be an enormous overlap between $M_a^\ell(S)$ and $M_n^\ell(S)$ when S is a k-space. We exploit this observation for the rest of this section. But first we describe $M_n^\ell(S)$ in the next Proposition. We leave the proof for the reader to supply.

2.5 PROPOSITION. (i) $M_n^\ell(S)$ *is a norm closed right ideal of* M(S).

(ii) $M_n(S)$ *is a closed subalgebra of* M(S).

Remark. Unfortunately in general even when S is locally compact the spaces $M_n^\ell(S)$ and $M_n(S)$ are not solid and so are not L-subalgebras. For an example illustrating this fact see Example 1.15(f). However we already know that $M_a^\ell(S)$ and $M_a(S)$ are L-subalgebras. This together with Corollary 2.4 suggests that $M_a(S)$ is a better behaved object than $M_n(S)$. So we shall formulate most of our results with reference to $M_a(S)$ instead of $M_n(S)$ even though some of these results are applicable to $M_n(S)$.

For a locally compact togological group G, as noted in Example 1.15(c), we have that $L(G,m) = L(G,m_1)$ for any left Haar measure m and right Haar measure m_1. The corresponding result for semigroups would be $M_a^\ell(S) = M_a^r(S)$.

In general one can easily formulate examples of semigroups S with $M_a^\ell(S) \neq M_a^r(S)$. However, for a large class of semigroups, we have the following affirmative result.

2.6 THEOREM. *Suppose* S *is a* k-*space with identity element* 1 *and* S *coincides with the foundation of* $M_a(S)$. *Then*

$$M_a^\ell(S) = M_a^r(S).$$

Proof. Let $\mu \in M_a^\ell(S)$ be fixed and (V_α) be a contracting neighbourhood basis of 1. We can find, (since $M_a(S)$ is solid), positive measures ν_α in $M_a(S)$ with $\text{supp}(\nu_\alpha) \subseteq V_\alpha$ and $\nu_\alpha(V_\alpha) = 1$. Hence, by Lemma 1.6(ii), we get

$$|h(\nu_\alpha * \mu) - h(\mu)| \leq \int |h(\bar{x} * \mu) - h(\mu)| d\nu_\alpha(x) \to 0$$

for all $h \in M(S)^*$. Hence μ is a weak cluster point of the net $(\nu_\alpha * \mu)$ in $M_a(S)$ (recall that $M_a(S)$ is an ideal), and so $\mu \in M_a(S)$. Hence $M_a^\ell(S) = M_a(S)$ and similarly $M_a^r(S) = M_a(S)$.

2.7 COROLLARY. *Suppose* S *is a* k-*space coinciding with the foundation of* $M_a(S)$ *and suppose* S *has an identity element* 1. *Then for every* $\mu \in M_a^\ell(S)$ *we have that*

(i) ℓ_μ *is norm continuous;*

(ii) r_μ *is norm continuous.*

Proof. Let $\mu \in M_a^\ell(S)$ and $(\nu_\alpha) \subset M_a(S)$ be chosen as in the proof of Theorem 2.6. Then $(\nu_\alpha * \mu) \subset M_n^\ell(S)$ by Corollary 2.4(ii). Now $M_n^\ell(S)$ is norm (and hence weakly) closed and μ the weak limit of $(\nu_\alpha * \mu)$. So $\mu \in M_n^\ell(S)$. Recalling Theorem 2.6 we similarly obtain $\mu \in M_n^r(S)$.

Both in Theorem 2.6 and Corollary 2.7, we have used the notion of an approximate identity without explicitly stating so. When translates of measures in $M_a(S)$ are weakly (or norm) continuous we naturally get situations where $M_a(S)$ has a weak (or norm, respectively) approximate identity. Before citing such situations we first examine this notion for general normed algebras for ease of reference. We follow Bonsall and Duncan [11] for the latter description.

2.8 DEFINITIONS. Let A be a (complex) normed algebra and (e_λ) a net in A.

(a) If $\text{norm-}\lim_\lambda e_\lambda x = x$ (or $\text{norm-}\lim_\lambda xe_\lambda = x$) for all $x \in A$, then the net (e_λ) is called a *left* (or *right*, respectively) *approximate identity* for A.

(b) If $\text{w-}\lim_\lambda e_\lambda x = x$ (or $\text{w-}\lim_\lambda xe_\lambda = x$) for all $x \in A$, then the net (e_λ) is called a *left* (or *right*, respectively) *weak approximate identity* for A.

(c) If there exists a positive constant c such that $\|e_\lambda\| < c$ for all λ, then the net (e_λ) is said to be *bounded* or to have *bound* c.

(d) *A bounded left* (*weak*) *approximate identity* is a left (weak) approximate identity which is also a bounded net. Similarly one defines *a bounded right* (*weak*) *approximate identity*. *A two-sided approximate identity* (or simply *an approximate identity*) is a net which is both a left and right approximate identity. Similarly one defines *a weak approximate identity*.

We now have the following result of Altman (see e.g. Bonsall and Duncan [11]).

2.9 PROPOSITION. *Let* A *be a normed algebra containing a bounded set* U *such that, given* $x \in A$ *and* $\varepsilon > 0$, *there exists* $u \in U$ *with* $\|x-ux\| < \varepsilon$. *Then* A *has a bounded left approximate identity.*

Proof. Let $\varepsilon > 0$ and $\{x_1,\dots,x_{n+1}\}$ be a finite subset of A. We shall prove that there exists $w \in W := \{vu : v,u \in U\}$ such that

$$\|x_i-wx_i\| < \varepsilon \quad \text{for all} \quad i \in \{1,\dots,n+1\}.$$

Then, in a standard way, one can construct a net in W which is a bounded left approximate identity for A.

Suppose $\|u\| \leq c$ for all $u \in U$. Choose $u,v \in U$ such that

$$\|x_1-ux_1\| < \frac{\varepsilon}{1+c}, \quad \|(x_2-ux_2) - v(x_2-ux_2)\| < \varepsilon.$$

With $w := vu$ we then have $\|x_i-wx_i\| < \varepsilon$ for $i = 1,2$. By the inductive hypothesis, suppose the result has been established for $\{x_1,\dots,x_n\}$. Set $k := \max\{\|x_i\| : i = 1,\dots,n\}$. There exists $y \in W$ with

$$\|x_i-yx_i\| < \frac{\varepsilon}{3(1+c)^2} \quad \text{for} \quad i = 1,\dots,n .$$

Now choose $w \in W$ such that

$$\|y-wy\| < \frac{\varepsilon}{3c} \quad \text{and} \quad \|x_{n+1}-wx_{n+1}\| < \varepsilon.$$

Then for $i = 1,\ldots,n$ we have that

$$\|x_i-wx_i\| \leq \|x_i-yx_i\| + \|y-wy\| \, \|x_i\| + \|w\| \, \|x_i-yx_i\| < \varepsilon.$$

This completes our proof.

2.10 PROPOSITION. *Let* A *be a normed algebra with a bounded left weak approximate identity. Then* A *has a bounded left approximate identity.*

Proof. Let (e_λ) be a bounded left weak approximate identity for A, and let U be the convex hull of $\{e_\lambda : \text{all } \lambda\}$. Since $\omega\text{-}\lim_\lambda e_\lambda x = x$ for all $x \in A$, there exists $u \in U$ with $\|x-ux\| < \varepsilon$. Then Proposition 2.9 implies the result.

2.11 COHEN'S FACTORIZATION THEOREM (see e.g. [11]). *Let* A *be a Banach algebra with a bounded left approximate identity. Then* AA = A.

2.11 APPROXIMATE IDENTITIES FOR $M_a(S)$. Let S be a k-space coinciding with the foundation of $M_a(S)$ and 1 be an identity element of S. Let (V_α) be a contracting net of neighbourhoods of 1. Since $M_a(S)$ is solid we can choose positive measures $\nu_\alpha \in M_a(S)$ with $\mathrm{supp}(\nu_\alpha) \subset V_\alpha$ and $\nu_\alpha(V_\alpha) = 1$. Then a glance at the proof of Theorem 2.6 teaches us that (ν_α) is a bounded weak approximate identity for $M_a(S)$. Recalling Proposition 2.9 we have that $M_a(S)$ has bounded approximate identity. (In fact by arguing as done by Sleijpen [92]) one can show that (ν_α) is a bounded approximate identity for $M_a(S)$.)

Similarly if 1 is a left identity of S and S is a k-space coinciding with the foundation of $M_a^\ell(S)$ then $M_a^\ell(S)$ has a bounded left approximate identity.

As noted in [92], the existence of a (bounded left approximate) identity for $M_a^\ell(S)$ (in fact for $\ell^1(S)$) does not in general imply that S has an identity element. For example consider $S := \{a,b,c\}$ with multiplication given by

$$aa := a, \quad bb := b, \quad ab = ba := c \quad \text{and} \quad c \text{ is a zero element.}$$

Then $\ell^1(S) = M(S)$ and $\nu := \bar{a} + \bar{b} - \bar{c}$ is the identity element of $M(S)$, but S does not have an identity element.

Recalling Cohen's factorization theorem we have

2.12 COROLLARY. *Let* S *be a* k-*space coinciding with the foundation of* $M_a(S)$ *and suppose that* S *has an identity element.* *Then* $M_a(S) * M_a(S) = M_a(S)$.

3. EQUI-REGULAR MEASURES

For any topological semigroup S we consider the spaces

$M_e^\ell(S) := \{\mu \in M(S)$: given $\varepsilon > 0$ and any compact $C \subset S$, there is an open neighbourhood W of C with $|\mu|(x^{-1}(W\setminus C)) < \varepsilon$ for all $x \in S\}$

$M_e^r(S) := \{\mu \in M(S)$: given $\varepsilon > 0$ and any compact $C \subset S$, there is an open neighbourhood W of C with $|\mu|((W\setminus C)x^{-1}) < \varepsilon$ for all $x \in S\}$

$M_e(S) := M_e^\ell(S) \cap M_e^r(S)$.

We then say $\mu \in M(S)$ is *left equi-regular* if $\mu \in M_e^\ell(S)$ and *equi-regular* if $\mu \in M_e(S)$. As will be apparent subsequently, these three subspaces of $M(S)$ can be identified with $L^1(S)$ if S is a locally compact topological group.

3.1 LEMMA. *Let* S *be a* C-*distinguished topological semigroup and* $\mu \in M(S)$. *Consider the following items:*

(i) $\mu \in M_e^\ell(S)$;

(ii) *given any compact subsets* D, F *and* K *of* S *and* $\varepsilon > 0$, *there is an open neighbourhood* W *of* K, *such that*

$$|\mu|(x^{-1}(W\setminus K)) < \varepsilon \textit{ for all } x \textit{ in } DF^{-1};$$

(iii) *the set* $\{(\bar{x}*|\mu|)_{|K} : x \in S\}$ *is relatively weakly compact, for all compact* $K \subset S$.

Then item (i) *implies* (ii) *and* (iii), *and if* S *is locally compact we have that these three items are equivalent.*

Proof. That (i) implies (ii) follows from the definition of $M_e^\ell(S)$. Again given a compact subset K of S, we have that $\mu \in M_e^\ell(S)$ implies that $A := \{(x*|\mu|)_{|K} : x \in S\}$ satisfies condition (ii) of Grothendieck's Theorem B4 (in Appendix) and, by the latter Theorem, A is relatively weakly compact. Thus (i) implies (iii).

Now suppose that S is locally compact for the remainder of this proof.

We assume that (iii) holds and prove (i). Let C be a given compact subset of S and V be a compact neighbourhood of C. Given $\varepsilon > 0$ and taking $A := \{(\bar{x}*|\mu|)_{|V} : x \in S\}$ in Appendix B.4, we have that item (iii) implies the existence of an open neighbourhood W_o of C with

$$\{(\bar{x}*|\mu|)_{|V}\,(W_o\setminus C) < \varepsilon \quad \text{for all } x \in S.$$

So if $W := \operatorname{int}(W_o \cap V)$, then W is an open neighbourhood of C and

$$\begin{aligned}|\mu|(x^{-1}(W\setminus C)) &= \bar{x}*|\mu|(W\setminus C) \cap V)\\ &= (\bar{x}*|\mu|)_{|V}\,(W\setminus C) < \varepsilon \quad \text{for all } x \in S.\end{aligned}$$

Thus $\mu \in M_e^{\ell}(S)$.

Finally we show that (ii) implies (i). Let $\varepsilon > 0$ be given. We can choose a compact subset F such that $\|\,|\mu|-|\mu|_{|F}\,\| < \varepsilon$. Let D be any compact neighbourhood of a given compact set $C \subset S$. Then (ii) implies the existence of an open set W with

$$C \subset W \subset D \quad \text{and} \quad |\mu|(x^{-1}(W\setminus C)) < \varepsilon \quad \text{for all } x \text{ in } DF^{-1}.$$

If $x \notin DF^{-1}$, then $x^{-1}D \cap F = \phi$ and hence $x^{-1}W \cap F = \phi$ and so

$$|\mu|(x^{-1}(W\setminus C)) = |\mu|(x^{-1}(W\setminus C)\setminus F) \le \|\,|\mu|-|\mu|_{|F}\| < \varepsilon.$$

Consequently,

$$|\mu|(x^{-1}(W\setminus C)) < \varepsilon \text{ for all } x \in S.$$

Thus (ii) implies (i).

3.2 THEOREM. *For any topological semigroup* S, *we have that*

(a) $M_e^{\ell}(S) \subseteq M_a^{\ell}(S)$.

(b) *If* S *is such that* CD^{-1} *is compact whenever* C *and* D *are compact subsets, then*

$$M_e^{\ell}(S) = M_a^{\ell}(S).$$

Proof. (a) Let $\mu \in M_e^{\ell}(S)$ be fixed. Let C be any compact subset of S and (x_α) be a net converging to x in S. By Lemma 1.4.1(ii), the mapping $s \to |\mu|(s^{-1}C)$ of S into $\mathbb{R}$ is continuous if

$$\varliminf_{\alpha} |\mu|(x_\alpha^{-1}C) \geq |\mu|(x^{-1}C).$$

Now let $\varepsilon > 0$ be given and choose an open neighbourhood W of C such that $|\mu|(s^{-1}(W\setminus C)) < \varepsilon$ for all s in S. We then have

$$\begin{aligned}\varliminf_{\alpha} |\mu|(x_\alpha^{-1}C) &= \varliminf_{\alpha}(|\mu|(x_\alpha^{-1}W) - |\mu|(x_\alpha^{-1}(W\setminus C)))\\ &\geq \varliminf_{\alpha} |\mu|(x_\alpha^{-1}W) - \varepsilon\\ &\geq |\mu|(x^{-1}W) - \varepsilon \qquad \text{by Lemma 1.4.1(i)}\\ &\geq |\mu|(x^{-1}C) - \varepsilon.\end{aligned}$$

Hence $\mu \in M_a^{\ell}(S)$.

(b) Let $\nu \in M_a^{\ell}(S)$ and D, F be any compact subsets of S. Then since DF^{-1} is compact, we have $\{\bar{x}*|\nu| : x \in DF^{-1}\}$ weakly compact, by Lemma 1.4. From Lemma 3.1, Corollary 1.10 and Appendix B.4 we have that $\nu \in M_e^{\ell}(S)$.

The preceding Theorem suggests that before studying the space $M_e(S)$ in detail it is essential to know whether $M_e(S)$ does not always coincide with $M_a(S)$. Our next Proposition will show that $M_e(S)$ is indeed generally very far from being equal to $M_a(S)$, even for locally compact topological semigroups S. In fact $M_e(S)$ is not even an L-ideal in general (see Proposition 3.9).

3.3 NOTATION. Whenever S is a topological semigroup, we shall denote the foundation of $M_a(S)$ by $F_a(S)$ and that of $M_e(S)$ by $F_e(S)$.

As remarked in Chapter 1 this notion of "foundation" was first used by A.C. and J.W. Baker with reference to $M_a(S)$. In particular following A.C. and J.W. Baker we shall say S is a *foundation semigroup* if S is a locally compact topological semigroup such that $F_a(S) = S$.

3.4 PROPOSITION. *There exists a commutative locally compact topological semigroup* S *such that* $F_a(S) = S$ *and* $F_e(S) = \phi$.

Proof. Let $S_n := \{(x,y) : x \in [2^n, 2^{n+1}-1]$ and $y = n-1$ or $n - 1 + \frac{1}{m}$ for some $m \in \mathbb{N}\} \cup \{(x,y) : x \in [2^{n+1}-1, 2^{n+1}]$ and $y \in [n-1,n]\}$ and $S := \bigcup_{n=1}^{\infty} S_n$. We now define the operation on S as follows: For any (x_1,y_1)

and (x_2,y_2) in S, there exists n_1 and n_2 with $(x_1,y_1) \in S_{n_1}$ and $(x_2,y_2) \in S_{n_2}$. We may suppose $n_1 \leq n_2$. Let

$$(x_2,y_2)(x_1,y_1) = (x_1,y_1)(x_2,y_2) := \begin{cases} (x_1+x_2, \max(y_1,y_2)) & \text{if } x_1 + x_2 \leq 2^{n_2} \\ (x_1+x_2, n_2) & \text{if } x_1 + x_2 > 2^{n_2}. \end{cases}$$

Next we give S the restriction topology of the plane.

(The reader may find it helpful to draw a sketch of this topological semigroup.)

Now one can easily verify that S is an abelian locally compact topological semigroup.

It is now a matter of routine to check that two-dimensional lebesgue measure on $\{(x,y) : x \in [2^{n+1}-1, 2^{n+1}]$ and $y \in [n-1, n]\}$, and for each $y = n-1 + \frac{1}{m}$ one-dimensional lebesgue measure on $\{(x,y) : x \in [2^n, 2^{n+1}-1]\}$ belong to $M_a(S)$, for all $m,n \in \mathbb{N}$. Consequently $F_a(S) = S$.

To show that $F_e(S) = \phi$ we consider a non-zero measure $\mu \in M(S)$. By the countable additivity of $|\mu|$, there is an n in $\mathbb{N}$ with $|\mu|(S_n)$ non-zero. (Let this n be fixed.) Since

$$S_n = \bigcup_{k=1}^{2^n} A_k \quad \text{where} \quad A_k := \{(x,y) \in S_n : x \in [2^n+(k-1), 2^n+k]\}$$

there is a $k_1 \in \{1,\ldots,2^n\}$ with $|\mu|(A_{k_1})$ non-zero. Since $M_e(S)$ is solid (see Thoerem 3.5), μ will not be an element of $M_e(S)$ if $\mu_{|A_{k_1}}$ is not in $M_e(S)$.

To prove that $\mu_{|A_{k_1}}$ is not in $M_e(S)$ let K be the compact set

$\{(x,n) : x \in [2^{n+2}-1, 2^{n+2}]\}$ and W any open neighbourhood of K. Then there exists $y_1 := n + \frac{1}{m}$ for some $m \in \mathbb{N}$ such that

$$\{(x,y_1) : x \in [2^{n+2}-1, 2^{n+2}]\} \subset W\backslash K.$$

Let $x_1 := 2^{n+2} - 2^n - k_1$ and observe that $x_1 \in (2^{n+1}, 2^{n+2})$, so (x_1,y_1) is an element of S_{n+1} (and hence of S). Taking $a := (x_1,y_1)$, we get $A_{k_1} \subset a^{-1}(W\backslash K)$. Hence

$$|\mu_{|A_{k_1}}|(a^{-1}(W\backslash K)) = \|\mu_{|A_{k_1}}\| .$$

So $\mu|_{A_{k_1}}$ is not a member of $M_e(S)$. As remarked before, this shows that every non-zero member of M(S) does not belong to $M_e(S)$. Consequently $F_e(S)$ is empty.

In our next two Theorems we describe the structure of the spaces of equiregular measures.

For the remainder of this section take S *to be a* C*-distinguished topological semigroup.*

3.5 THEOREM. $M_e^\ell(S)$ *is a left* L*-ideal of* M(S).

Proof. Let $\mu \in M_e^\ell(S)$ and $\nu \in M(S)$ with $\nu << |\mu|$. By the Radon-Nikodym theorem given $\varepsilon > 0$, there exists a $\delta > 0$ such that $|\mu|(E) < \delta$ implies $|\nu|(E) < \varepsilon$ for all Borel $E \subset S$. Hence if C is any compact subset and W an open neighbourhood of C in S such that $|\mu|(x^{-1}(W\setminus C)) < \delta$ for all x in S, we get $|\nu|(x^{-1}(W\setminus C)) < \varepsilon$ for all x in S and $\nu \in M_e^\ell(S)$.

Now to show that $M_e^\ell(S)$ is closed it is sufficient to consider (a positive measure) $\mu \in M(S)$ such that $\|\mu_n-\mu\| \to 0$ for some sequence of positive measures $\{\mu_n\} \subseteq M_e^\ell(S)$. Let C be any compact subset of S and $\varepsilon > 0$ be given. There is $n' \in \mathbb{N}$ such that $\|\mu-\mu_{n'}\| < \frac{\varepsilon}{2}$. Choose an open neighbourhood W of C such that $\mu_{n'}(x^{-1}(W\setminus C)) < \frac{\varepsilon}{n}$. Then for all $x \in S$ we get

$$\mu(x^{-1}(W\setminus C)) = (\mu-\mu_{n'})(x^{-1}(W\setminus C)) + \mu_{n'}(x^{-1}(W\setminus C))$$

$$\leq \|\mu-\mu_{n'}\| + \mu_{n'}(x^{-1}(W\setminus C)) < \varepsilon.$$

So $\mu \in M_e^\ell(S)$ and $M_e^\ell(S)$ is closed.

To complete our proof it is sufficient to prove that $M_e^\ell(S)$ is a left ideal. To the latter end we fix $\mu \in M_e^\ell(S)$ and $\nu \in M(S)$. Let C be a compact subset of S and $\varepsilon > 0$ be given. We can find an open neighbourhood W of C such that

$$|\mu|(x^{-1}(W\setminus C)) < \frac{\varepsilon}{(1+\|\nu\|)}$$

for all $x \in S$. Hence, for all $x \in S$, we have that

$$|\nu|*|\mu|(x^{-1}(W\setminus C)) = \int |\mu|(y^{-1}x^{-1}(W\setminus C))d|\nu|(y)$$

$$= \int |\mu|(xy)^{-1}(W\setminus C)d|\nu|(y)$$

$$\leq \frac{\varepsilon}{(1+\|\nu\|)}\|\nu\| < \varepsilon .$$

So $|\nu|*|\mu| \in M_e^{\ell}(S)$. Since $\nu*\mu << |\nu|*|\mu|$, it follows that $\nu*\mu \in M_e^{\ell}(S)$.

3.6 LEMMA. *Let* $\mu \in M_e^{\ell}(S)$ *and* K *be a compact subset of* S. *Then the map* $x \to (\bar{x}*\mu)_{|K}$ *of* S *into* M(S) *is weakly continuous.*

Proof. Let (x_α) be a net converging to x in S. Noting that $(\bar{s}*\mu)_{|K} << (\bar{s}*|\mu|)_{|K}$ for all s in S, we have that the net $((\bar{x}_\alpha*\mu)_{|K})$ is relatively weakly compact by Lemma 3.1, and so must have a weak cluster point η (say). Now for every Borel set E, Theorem 3.2(a) and Lemma 1.2 imply that

$$(\bar{x}_\alpha*\mu)_{|K}(E) = \bar{x}_\alpha*\mu(E \cap K) \to \bar{x}*\mu(E \cap K) = (\bar{x}*\mu)_{|K}(E).$$

Consequently $(\bar{x}*\mu)_{|K} = \eta$ and $(\bar{x}*\mu)_{|K}$ is the only weak cluster point of $((\bar{x}_\alpha*\mu)_{|K})$. This completes our proof.

3.7 THEOREM. *If* S *is such that* $F_e(S) = S$, *then* $M_e(S)$ *is an* L-*ideal of* M(S).

Proof. In view of Theorem 3.5, it follows that $M_e(S)$ is an L-subalgebra. We now show that $M_e(S)$ is left translation invariant. Let $\mu \in M_e(S)$ be fixed ν be a positive measure in $M_e(S)$ with D := supp(ν) compact. Fix $x \in D$ and let (V_α) be a net of open neighbourhoods contracting to x. Setting

$$\nu_\alpha := \|\rho_\alpha\|^{-1}\rho_\alpha \quad \text{where } \rho_\alpha := \nu_{|D\cap V_\alpha}$$

we have $(\nu_\alpha) \subseteq M_e(S)$ and $\|\nu_\alpha\| = 1$. Since $M_e(S) \subseteq M_a^{\ell}(S)$, Lemmas 1.4 and 1.6 imply that

$$|h(\nu_\alpha*\mu) - h(\bar{x}*\mu)| = |\int(h(\bar{s}*\mu) - h(\bar{x}*\mu))d\nu_\alpha(s)|$$

$$\leq \int |h(\bar{s}*\mu) - h(\bar{x}*\mu)|d\nu_\alpha(s) \to 0,$$

for all $h \in M(S)^*$. Consequently $\bar{x}*\mu \in M_e(S)$, for all $x \in D$. Now the inner regularity of measures implies that $F_e(S)$ is the closure of $H := \cup\{\mathrm{supp}(\nu)$: ν is a positive measure in $M_e(S)$ with compact support$\}$. Thus for any $y \in S$, (since $F_e(S) = S$) we can find $(y_\gamma) \subset H$ converging to y. Then by Lemma 3.6 we have $\text{weak-}\lim_\gamma (\bar{y}_\gamma * \mu)_{|K} = (\bar{y}*\mu)_{|K}$ for any compact $K \subset S$. Since $(\bar{y}_\gamma * \mu)_{|K} \in M_e(S)$ we have $(\bar{y}*\mu)_{|K} \in M_e(S)$. Since $M_e(S)$ is closed and we can find a sequence of compact sets $\{K_n\}$ such that $\|(\bar{y}*\mu)_{|K_n} - \bar{y}*\mu\| \to 0$, we conclude that $\bar{y}*\mu \in M_e(S)$, for all $y \in S$.

So $M_e(S)$ is left translation invariant and (similarly) right translation invariant. Since $M_e(S) \subseteq M_a(S)$ one can now easily imitate the proof of Theorem 1.7 to show that $M_e(S)$ is an ideal of $M(S)$.

Other situations where $M_e(S)$ is an L-ideal can be found in Dzinotyiweyi [32] and Dzinotyiweyi and Sleijpen [39]. Unfortunately $M_e(S)$ is not always an ideal (or an L-ideal) of $M(S)$. To show this we use an example given in [39] for a situation where S is quite "nice" with respect to $M_a(S)$.

3.9 PROPOSITION. *There exists a locally compact topological semigroup* S *with* $F_a(S) = S$ *and such that* $M_e(S)$ *is non-zero but is neither a left nor right ideal of* $M(S)$.

Proof. Let $T_1 := [0,1] \times \{0\} \cup [0,1] \times \{1\} \cup \{\alpha\}$, where $\alpha := (1,2)$, be such that $[0,1] \times \{0\}$ has the discrete topology and $[0,1] \times \{1\} \cup \{\alpha\}$ the restriction topology of the usual topology on $\mathbb{R}^2$. T_1 is a locally compact topological semigroup under the multiplication defined as follows:

$$\left.\begin{array}{lll} (t,t')\alpha := \alpha(t,t') := \alpha\alpha := \alpha & & \\ (t,t')(s,s') := (t,s') & \text{if} & t'=0 \\ (t,t')(s,s') := \alpha & \text{if} & t'\neq 0 \end{array}\right\} \text{ for all } (t,t'),(s,s') \in T_1\backslash\{\alpha\} .$$

Taking $\beta := (1,3)$ we extend the multiplication on T_1 to $T := T_1 \cup \{\beta\}$ by

$$\left.\begin{array}{ll} \beta\beta := \beta,\ \beta\alpha := \alpha\beta := \alpha & \\ \beta(t,t') := (0,t') & \\ (t,t')\beta := (t,0) & \text{if } t'=0 \\ (t,t')\beta := \alpha & \text{if } t'\neq 0 \end{array}\right\} \text{ for all } (t,t') \in T\backslash\{\alpha\}.$$

Then T is a locally compact topological semigroup, (we omit the technical details for verifying this). Further it is not difficult to show that $F_a(T) = T$, (actually $\bar{\alpha}, \bar{\beta}, \bar{x} \in M_a(S)$ for all $x \in [0,1] \times \{0\}$ and lebesgue measure on the line $[0,1] \times \{1\}$ belongs to $M_a(T)$).

Now we have that

$$\left.\begin{aligned} \{\bar{\beta} * \bar{x} : x \in T\} &= \{\bar{\alpha}, \bar{\beta}, \overline{(0,1)}, \overline{(0,0)}\} \\ \{\bar{x} * \bar{\beta} : x \in T\} &= \{\bar{\alpha}, \bar{\beta}\} \cup \{\bar{x} : x \in [0,1] \times \{0\}\} \end{aligned}\right\} \quad (1)$$

Thus restrictions of sets in (1) to compact subsets of T yield finite subsets of M(T) and so, recalling Lemma 3.1, we conclude that $\bar{\beta} \in M_e(T)$.

For convenience let $\gamma := (0,1)$. Now

$$\{\bar{x} * \bar{\gamma} : x \in [0,1] \times \{0\}\} = \{\bar{x} : x \in [0,1] \times \{1\}\}$$

and so (again by Lemma 3.1) we have that $\bar{\gamma} \notin M_e(T)$. But

$$\bar{\beta} * \bar{\gamma} = \bar{\gamma}.$$

Hence $M_e(T)$ is not a right ideal of M(T).

Let b be an element isolated from $T \times \{1,2\}$ and set $S := T \times \{1,2\} \cup \{b\}$. We define multiplication on S as follows:

$$\left.\begin{aligned} sb := bs &:= b \quad \text{for all } s \in S \\ (x,1)(y,2) &:= (y,2)(x,1) := b \\ (x,1)(y,1) &:= (xy, 1) \\ (x,2)(y,2) &:= (yx, 2) \end{aligned}\right\} \quad \text{for all } x,y \in T.$$

Again the reader is left with the exercise of showing that S is a locally compact topological semigroup with $F_a(S) = S$.

Let $p := (\beta,1)$ and $q := (\beta,2)$ and note that $\bar{p}, \bar{q} \in M_e(S)$. However, taking $x_1 := (\gamma,1)$ and $x_2 := (\gamma,2)$ we have that $\bar{x}_1, \bar{x}_2 \notin M_e(S)$ (by using the fact that $\bar{\gamma} \notin M_e(T)$),

$$\bar{p} * \bar{x}_1 = \bar{x}_1 \quad \text{and} \quad \bar{x}_2 * \bar{q} = \bar{x}_2 .$$

Hence $M_e(S)$ is neither a left nor right ideal of M(S).

We conclude this section with a collection of Exercises for the reader.

3.10 EXERCISES.

(a) Let S be any topological semigroup and $x \in S$.

(i) If $xS \cup Sx$ is finite then $\bar{x} \in M_e(S)$;

(ii) if S is compact then $xS \cup Sx$ is finite if and only if $\bar{x} \in M_a(S)$.

(b) Let $S := [0,1]$ with the relative topology of the line and maximum operation. If $x \in S$, then $\bar{x} \in M_a(S)$ if and only if $x = 1$. (In fact every non-zero element of $M_a(S)$ is a scalar multiple of $\bar{1}$.)

(c) Let S be any topological semigroup and $x \in S$. We have

(i) if $\bar{x} \in M_a^{\ell}(S)$, then $\{sx\}x^{-1}$ is a neighbourhood of s, for all $s \in S$;

(ii) if x is a right cancellation element (e.g. if x is an identity element), then $\bar{x} \in M_a^{\ell}(S)$ if and only if S is discrete. [Hint: Recall Lemma 1.9.]

(d) Let $S := \{0\} \cup \{\frac{1}{n} : n \in \mathbb{N}\}$ with the relative topology of the line and multiplication given by $xy := \max(x,y)$ for all $x,y \in S$. Then S is a foundation semigroup such that for each $x \in S$ we have $\bar{x} \in M_a(S)$ if and only if $x \neq 0$.

(e) If S is a discrete topological semigroup, then $M_e^{\ell}(S) = M_e(S) = \ell^1(S)$.

(f) (i) Let S be a C-distinguished topological semigroup with a right cancellation element x. Whenever $\bar{x}$ is right equi-absolutely continuous we have $M(S) = \ell^1(S)$; and if, in addition, S is locally compact then S is discrete.

(ii) If S is as in (b), (d) or (e), we have that every measure $\mu \in M(S)$ is absolutely continuous if and only if μ is equi-absolutely continuous. In Chapter 5, section 2, we shall study the relationship between equi-absolutely continuous and absolutely continuous measures for *many* topological semigroups.)

(g) If S and T are foundation semigroups, then the product topological semigroup $S \times T$ is a foundation semigroup. (See e.g. [5].)

4. QUASI-INVARIANT MEASURES

Let S be a C-distinguished topological semigroup throughout this section.

A natural problem to ask is: Given that $M_a(S)$ (or $M_e(S)$) is non-trivial can one find a measure m on S, with some form of invariance property related to that of Haar measure on a topological group, and then characterize $M_a(S)$ (or $M_e(S)$, respectively) in terms of $L(S,m)$?

There does not seem to be a complete answer to such a problem. However the idea of quasi-invariant measures studied by e.g. Rothman [87] seems to be of interest in this direction. We will use such "quasi-invariant" measures to obtain some partial answer to the preceding problem in (the form of) Theorems 4.2 and 4.3.

4.1 DEFINITIONS. Let H be a Borel subset of S. A Radon measure m on S is said to be *left equi-quasi-invariant* on H if for every Borel set $B \subset H$ we have that

$$m(x^{-1}B) = 0 \quad \text{for all} \quad x \in S \quad \text{whenever} \quad |m|(B) = 0.$$

We denote the set of all measures in M(S) which are left equi-quasi-invariant on H by $M^{\ell}_{eq}(S,H)$.

A Radon measure m on S is said to be *left quasi-invariant* if it is left equi-quasi-invariant on S. We denote the set of all left quasi-invariant measures in M(S) by $M^{\ell}_{q}(S)$.

Recalling the definition of $F_e(S)$ from item 3.3 we have

4.2 THEOREM. *Let H be any σ-compact subset of S and suppose that* $F_e(S) = S$. *Then*

(i) $M_e(S) = \cup\{L(S,m) : m \in M_e(S)^+ \cap M^{\ell}_{eq}(S,H)\}$;

(ii) *if S is σ-compact, we have that*

$$M_e(S) = \cup\{L(S,m) : m \in M_e(S)^+ \cap M^{\ell}_{q}(S)\}.$$

Proof. Since $M_e(S)$ is an L-subalgebra to prove item (i), it is sufficient to show that corresponding to each measure μ in $M_e(S)$ there is a positive measure m in $M_e(S) \cap M^{\ell}_{eq}(S,H)$ such that $\mu \ll m$. To this end, let $\mu \in M_e(S)$ be given and suppose $H = \bigcup_{p=1}^{\infty} K_p$ where each K_p is compact. By Lemma 3.1, the set $\{(\bar{x}*|\mu|)_{|K_p} : x \in S\}$ is relatively weakly compact. So $\{(\bar{x}*|\mu|)_{|K_p} : x \in S\}$ satisfies Appendix B.4 item (iii). Hence

$$\left.\begin{array}{l}\text{there is a sequence } \{(\bar{x}_{n,p}*|\mu|)_{|K_p}\}_{n\in\mathbb{N}} \text{ such that if} \\ B \text{ is a Borel subset of } S \text{ with } (\bar{x}_{n,p}*|\mu|)_{|K_p}(B) = 0 \text{ for all} \\ n \in \mathbb{N}\text{, then } (\bar{x}*|\mu|)_{|K_p}(B) = 0 \text{ for all } x \in S.\end{array}\right\} \quad (*)$$

Setting $m := |\mu| + \sum_{1\le n,p<\infty} 2^{-(n+p)}\bar{x}_{n,p}*|\mu|$ we have that $m \in M_e(S)^+$ since $M_e(S)$ is an L-ideal (by Theorem 3.7), and $\mu << m$. So it remains only to show that $m \in M^{\ell}_{eq}(S,H)$. Let B be a Borel set contained in H and such that $m(B) = 0$. Then $\bar{x}_{n,p}*|\mu|\ (B \cap K_p) = 0$ for all $n \in \mathbb{N}$ and hence $\bar{x}*|\mu|(B \cap K_p) = 0$ for all $x \in S$, by (*). Hence $\bar{x}*|\mu|(B) = 0$ for all $x \in S$, since $B \subset \bigcup_{p=1}^{\infty} K_p$. In particular we have that

$$\bar{x}*(\bar{x}_{n,p}*|\mu|)(B) = \overline{xx}_{n,p}*|\mu|(B) = 0$$

for all $x \in S$ and $n,p \in \mathbb{N}$. Consequently

$$m(x^{-1}B) = \bar{x}*|\mu|(B) + \sum_{1\le n,p<\infty} 2^{-(n+p)}\ \bar{x}*(\bar{x}_{n,p}*|\mu|)(B)$$

$$= 0.$$

This completes our proof of item (i).

Item (ii) follows from item (i) by taking $H = S$.

In a related manner one can prove the following result.

4.3 THEOREM. *If* S *is σ-compact and* $F_a(S) = S$ *then*

$$M_a(S) = \cup\{L(S,m) : m \in M_a(S)^+ \cap M^{\ell}_q(S)\}.$$

(We omit the details of the proof as they are also contained in the proof of Theorem 5.4.2.)

4.4 REMARK. It is interesting to note that Theorem 4.2(i) does not hold if we replace $F_e(S)$ by $F_a(S)$ and $M_e(S)$ by $M_a(S)$, even when S is also locally compact and commutative.

For an example we take S to be the topological semigroup constructed in item 3.4. Now let μ be a lebesgue measure with $\text{supp}(\mu) = \{(x,1) : x \in [2,3]\}$ and let H be the compact subset $\{(x,y) : x \in [7,8] \text{ and } y \in [1,2]\}$. Then $|\mu| \in M_a(S)$. Suppose (on the contrary) there is a measure m in $M_a(S) \cap M^{\ell}_{eq}(S,H)$ with $|\mu| << m$. Then for some $m \in \mathbb{N}$ taking $y := 1 + \frac{1}{m}$, we must have $m(B_y) = 0$ where $B_y := \{(x,y) : x \in [7,8]\}$. But if $z := (s,y)$, then $|\mu|(z^{-1}B_y) = \|\mu\| > 0$ and so $m(z^{-1}B_y) \neq 0$. By this contradiction our assertion follows.

4.5. One may ask whether $M^{\ell}_{eq}(S,H)$ is contained in $M_e(S)$ or $M_a(S)$ when H has non-empty interior. Unfortunately this is not the case even when S is a subsemigroup of a group. (For example let S be the additive subsemigroup $[0,\infty)$ of $\mathbb{R}$ and take m to be the point mass at 4 and $H := [0,3]$. Then $m \in M^{\ell}_{eq}(S,H)$ but $m \notin M_a(S)$.) However we have the following exercise for the reader.

EXERCISE. Let S be a locally compact topological group with a Haar measure m. If H is a Borel subset of S with non-empty interior and μ is a measure in $M^{\ell}_{eq}(S,H)$, then $\mu << m$.

4.6 EXAMPLE. One can find a locally compact commutative topological semigroup S supporting a measure $m \in M_q(S)$ such that $\text{supp}(m) = S$ but $F_a(S) = \phi$ (i.e. $M_a(S)$ is zero). For example take

$$S := \{0\} \cup \{n+2^{-k} : n,k \in \mathbb{N} \cup \{0\}\}$$

with the relative topology of the line and multipication given by

$$xy := \max(x,y) \quad \text{for all } x,y \in S.$$

Let $\{x_n\}$ be an enumeration of S and take $m := \sum_{n=1}^{\infty} 2^{-n}\, \bar{x}_n$. Then $\text{supp}(m) = S$ and m is (trivially) left quasi-invariant.

Now by considering various compact subsemigroups of S and recalling Exercise 3.10(a)(ii) the reader can easily note that $M_a(S)$ is zero.

The study of absolute continuity of quasi-invariant measures on topological semigroups is postponed to Section 4 of Chapter 5.

5. MULTIPLICATIVE LINEAR FUNCTIONALS AND SEMICHARACTERS

Throughout this section, unless otherwise stated, S *denotes a commutative* C-*distinguished topological semigroup with* $F_a(S) = S$.

We present a theory for semigroups that is well known and well established in the case of groups. Though our results are formulated in terms of the algebra of absolutely continuous measures, the reader can easily note that they also can be formulated in terms of the algebra of equi-regular measures.

5.1 NOTATIONS. By a *semicharacter* on S we mean a non-zero function χ in m(S) such that $\chi(xy) = \chi(x)\chi(y)$ for all x,y in S. (One can easily note that $\|\chi\|_S \leq 1$.) We denote the set of all continuous semicharacters on S by $S^\wedge$.

Let A be a closed subalgebra of M(S). By *a multiplicative linear functional* on A we mean a non-zero functional $h \in A^*$ such that

$$h(\nu*\mu) = h(\nu)h(\mu) \quad \text{for all } \nu,\mu \in A .$$

The set of all multiplicative linear functionals on A is denoted by $A^\wedge$.

Our first interest is to establish a correspondence between $S^\wedge$ and $M_a(S)^\wedge$ (see Theorems 5.3 and 5.6). Towards this end, we first prove a lemma. (Recall the definition of $F_a(S)$ from item 3.3.)

5.2 LEMMA. *Let* μ *be a measure in* $M_a(S)$ *and* h *an element of* $M_a(S)^\wedge$. *Then* $x \to h(\bar{x}*\mu)$ *defines a continuous function on* S.

Proof. Since h is non-zero there is a measure ν in $M_a(S)$ such that $h(\nu) \neq 0$. Now $h(\bar{x}*\mu)h(\nu) = h(\bar{x}*\mu*\nu) = h(\bar{x}*\nu*\mu) = h(\bar{x}*\nu)h(\mu)$ for all x in S, so if $h(\mu) = 0$ we get $h(\bar{x}*\mu) = 0$ for all x in S and our result follows trivially.

Suppose $h(\mu) \neq 0$. By the inner regularity of μ we may assume that $D := \mathrm{supp}(\mu)$ is compact. The function $x \to h(\bar{x}*\mu)$ is continuous on DD, by Lemma 1.4, and clearly $\mathrm{supp}(\mu*\mu) \subseteq DD$. Now by Lemma 1.6(ii)

$$\int h(\bar{x}*\mu)\,d\mu*\mu(x) = h(\mu*\mu*\mu) = h(\mu)^3 \neq 0.$$

Hence, by Theorem 1.14(b), we can find $u \in DD \cap S_r$ such that $h(\bar{u}*\mu) \neq 0$. (Recall the definition of S_r from item 1.11.)

So for all $x \in S$ we have that

$$h(\bar{x}*\mu) = \frac{h(\bar{x}*\mu)h(\bar{u}*\mu)}{h(\bar{u}*\mu)} = \frac{h(\overline{xu}*\mu)h}{h(\bar{u}*\mu)}$$

and $x \to h(\overline{xu}*\mu)$ defines a continuous function on S, by definition of S_r and Lemma 1.4. Then the preceding equation implies our result.

5.3 THEOREM. *There exists a one-to-one mapping* τ *of* $M_a(S)^\wedge$ *onto* $S^\wedge$ *such that*

$$h(\mu) = \int \chi_h(x)d\mu(x)$$

for all $h \in M_a(S)^\wedge$ *where* $\tau(h) = \chi_h$ *is in* $S^\wedge$.

Proof. Let h be an element of $M_a(S)^\wedge$ and choose μ in $M_a(S)$ with $h(\mu) \neq 0$. We define the function χ_h on S by $\chi_h(x) := h(\bar{x}*\mu)/h(\mu)$ for all x in S. Then $\chi_h \in C(S)$ by Lemma 5.2. (One can easily note that χ_h is independent of $\mu \in M_a(S)$ with $h(\mu) \neq 0$.) Now for all x,y in S we have

$$\chi_h(xy) = \frac{h(\overline{xy}*\mu)h(\mu)}{h(\mu)\ h(\mu)} = \frac{h(\bar{x}*\mu*\bar{y}*\mu)}{h(\mu)\ h(\mu)} = \frac{h(\bar{x}*\mu)}{h(\mu)}\frac{h(\bar{y}*\mu)}{h(\mu)} = \chi_h(x)\chi_h(y).$$

Further, by Lemma 1.6, we have

$$0 \neq h(\mu)^2 = h(\mu*\mu) = \int h(\bar{x}*\mu)d\mu(x)$$

and so the function $x \to h(\bar{x}*\mu)$ is non-zero. Hence $\chi_h \in S^\wedge$.

We now set

$$\tau(h) := \chi_h.$$

Clearly τ is a well-defined mapping of $M_a(S)^\wedge$ into $S^\wedge$.

To see that τ is onto and one-to-one fix $\chi \in S^\wedge$ and, corresponding to this χ, define $h \in M_a(S)*$ by $h(\nu) := \int \chi(x)d\nu(x)$. Since χ is non-zero and continuous, and $M_a(S)$ is solid, we certainly have that h is non-zero. Further

$$h(\mu)\chi(x) = \int \chi(x)\chi(y)d\mu(y) = \int \chi(xy)d\mu(y)$$

$$= \int \chi(y)d\bar{x}*\mu(y) = h(\bar{x}*\mu) = h(\mu)\ \frac{h(\bar{x}*\mu)}{h(\mu)}$$

$$= h(\mu)\ \chi_h(x)$$

and so $\chi = \chi_h$. From this it is now trivially to note that τ is both onto and one-to-one. Further the equation $h(\mu) = \int \chi_h(x)d\mu(x)$ stated in our Theorem is also now evident.

The preceding proof is very much alike to that used in the group case - see e.g. Hewitt and Ross [59].

5.4 COROLLARY. *Let* χ *be a semicharacter on* S *that is* $M_a(S)$*-measurable and such that the set* $\{x \in S : \chi(x) \neq 0\}$ *is not* $M_a(S)$*-negligible. Then* χ *is continuous* (i.e. $\chi \in S^\wedge$).

Proof. We can define $h \in M_a(S)^\wedge$ by $h(\mu) := \int \chi(x)d\mu(x)$ for all $\mu \in M_a(S)$ and hence $\chi_h \in S^\wedge$ as done in proof of the preceding Theorem. Now arguing as in the last paragraph of the proof of Theorem 5.3 we get $\chi = \chi_h$.

5.5 DEFINITIONS. For all μ in $M_a(S)$, the *Fourier transform* $\hat{\mu}$ is defined by $\hat{\mu}(h) := h(\mu)$ for all $h \in M_a(S)^\wedge$. Note that $\hat{\mu}$ is a function on $M_a(S)^\wedge$. The *Gel'fand topology* on $M_a(S)^\wedge$ is the weakest topology on $M_a(S)^\wedge$ under which all of the functions $\hat{\mu}$ are continuous. It follows trivially from a more general result (see e.g. Hewitt and Ross [59, Appendix C.26]) that $M_a(S)^\wedge$, with the Gel'fand topology, is a locally compact space.

We define the *compact open topology* on C(S) as follows: a net (f_α) converges to the zero function in C(S) in the compact open topology if and only if $\|\psi f_\alpha\|_S \to 0$ for every $\psi \in m(S)$ with compact support.

With these definitions in mind we have

5.6 THEOREM. *If in Theorem* 5.3 *the set* $M_a(S)^\wedge$ *is endowed with the Gel'fand topology and* $S^\wedge$ *with the compact open topology, then the mapping* τ *is a homeomorphism.*

In particular if in addition S *has an identity element then* $S^\wedge$ *(with the compact open topology) is a locally compact topological semigroup under pointwise multiplication.*

Proof. First we show that τ is an open map. Let (χ_α) be a net converging to χ_o in $S^\wedge$ with the compact open topology and set $h_\alpha := \tau^{-1}(\chi_\alpha)$. Let μ be any element of $M_a(S)$ and $\varepsilon > 0$ be given. We can find a compact set $K \subseteq S$ such that $\|\mu - \mu_{|K}\| < \frac{\varepsilon}{4}$. Now eventually

$$|\chi_\alpha(x) - \chi_o(x)| < \frac{\varepsilon}{2(1+|\mu|(K))}$$

for all $x \in K$. Hence, eventually,

$$\begin{aligned}|\hat{\mu}(h_\alpha) - \hat{\mu}(h_o)| &= \left|\int (\chi_\alpha(x) - \chi_o(x))\,d\mu(x)\right| \\ &\leq \int_K |\chi_\alpha(x) - \chi_o(x)|\; d|\mu|(x) + 2\,\|\mu - \mu_{|K}\| \\ &< \varepsilon.\end{aligned}$$

Thus τ is an open map.

Next we show that τ is continuous. Suppose, on the contrary, τ is not continuous. Then for some $\varepsilon > 0$ and compact set $K \subseteq S$, one can find a net (h_β) converging to h in $M_a(S)^\wedge$ (endowed with the Gel'fand topology) and (x_β) contained in K, such that

$$|\chi_{h_\beta}(x_\beta) - \chi_h(x_\beta)| \geq \varepsilon \qquad (*)$$

for all β. Since K is compact we may assume that (x_β) converges to some point x in K. We can find $\mu \in M_a(S)$ with $h(\mu) \neq 0$. Since $\hat{\mu}(h_\beta) \to \hat{\mu}(h)$, we have that $h_\beta(\mu) \neq 0$ eventually. So, eventually, we have that

$$\begin{aligned}|\chi_{h_\beta}(x_\beta) - \chi_h(x_\beta)| &\leq \left|\frac{h_\beta(\bar{x}_\beta*\mu)}{h_\beta(\mu)} - \frac{h_\beta(\bar{x}*\mu)}{h_\beta(\mu)}\right| + \left|\frac{h_\beta(\bar{x}*\mu)}{h_\beta(\mu)} - \frac{h(\bar{x}_\beta*\mu)}{h(\mu)}\right| \\ &= \left|\frac{1}{(h_\beta(\mu))^2}\, h_\mu(\bar{x}_\beta*\mu*\mu - \bar{x}*\mu*\mu)\right| + \left|\frac{h_\beta(\bar{x}*\mu)}{h_\beta(\mu)} - \frac{h(\bar{x}_\beta*\mu)}{h(\mu)}\right| \\ &\leq \frac{1}{|h_\beta(\mu)|^2}\, \|\bar{x}_\beta*\mu*\mu - \bar{x}*\mu*\mu\| + \left|\frac{h_\beta(\bar{x}*\mu)}{h_\beta(\mu)} - \frac{h(\bar{x}_\beta*\mu)}{h(\mu)}\right| \\ &\to 0,\end{aligned}$$

by Theorem 2.3. This contradicts $(*)$, and hence τ is continuous.

It follows that τ is a homeomorphism.

In particular $\hat{S}$ is therefore a locally compact space, by the local compactness of the Gel'fand topology. If in addition S has an identity element 1 (say), then for any $\chi,\psi \in \hat{S}$ we have $\chi.\psi(1) := \chi(1)\psi(1) = 1$ and one can easily note that $\chi.\psi \in \hat{S}$, and hence that $\hat{S}$ is a topological semi-group.

5.7 COROLLARY. *If* S *has an identity element* 1 *then the following items are equivalent:*

(i) $\hat{S}$ *(endowed with the compact open topology) is compact;*

(ii) S *is discrete.*

Proof. Item (i) implies that $M_a(S)^\wedge$ is compact, by Theorem 5.6. Now the compactness of $M_a(S)^\wedge$ in turn implies that $\bar{1} \in M_a(S)$, (see e.g. Hewitt and Ross [59, Appendices C.24 and C.25]). Now from Exercises 3.10(c)(ii), we have that S is discrete.

That item (ii) implies (i) follows trivially from the definitions.

We now turn to the question of semi-simplicity of the algebra $M_a(S)$.

5.8. We recall some standard results for a commutative Banach algebra A. Let $\hat{A} := \{h \in A^* : h(ab) = h(a)h(b)$ for all $a,b \in A$ and h is non-zero$\}$.

Every ideal I of A is maximal if and only if I is the kernel of some element of $\hat{A}$. The *radical* of A, denoted by rad(A), is the intersection of all maximal ideals of A. (We assume this definition also for cases where A is not commutative.) A is said to be *semi-simple* if $\text{rad}(A) = \{0\}$. Thus, equivalently, A is semi-simple if and only if for every $a \in A$ with $h(a) = 0$, for all $h \in \hat{A}$, we must have a equal to the zero element of A. We shall use the latter observation in the proof of our next Theorem.

5.9 THEOREM. *Let* S *have an identity element* 1. *Then the following items are equivalent.*

(i) $M_a(S)$ *is semisimple,*

(ii) M(S) *is semisimple,*

(iii) $\hat{S}$ *separates the points of* S.

Proof. Suppose item (i) holds. We prove that item (ii) holds. Let $\mu,\nu \in M(S)$ be distinct. Recalling that $M(S)$ can be identified with $C_\beta(S)^*$ we can find $f \in C(S)$ such that $\mu(f) \neq \nu(f)$. Thus we have $\mu of, \nu of \in C(S)$ with

$$\mu of(1) = \mu(f) \neq \nu(f) = \nu of(1).$$

So Proposition 1.3.3 implies that we can find $\rho \in M_a(S)$ such that

$$\mu*\rho(f) - \nu*\rho(f) = \int (\mu of - \nu of)(x)d\rho(x) \neq 0.$$

Thus $\mu*\rho \neq \nu*\rho$ and we can find $h \in M_a(S)^\wedge$ with $h(\mu*\rho) \neq h(\nu*\rho)$. In particular $h(\rho) \neq 0$, so we can extend h to $h' \in M(S)$ by setting

$$h'(\eta) := \frac{h(\eta*\rho)}{h(\rho)} \quad (\eta \in M(S)).$$

Clearly $h'(\mu) \neq h'(\nu)$ and so item (ii) holds.

Next we assume item (ii) and prove item (iii). Let x,y be distinct points of S. We can find an open neighbourhood V of 1 with $Vx \cap Vy = \phi$ and a non-zero measure $\nu \in M_a(S)$ with $\mathrm{supp}(\nu) \subset V$. Hence $\nu*\bar{x} \neq \nu*\bar{y}$ and we can find $h \in M(S)^\wedge$ with $h(\nu*\bar{x}) \neq h(\nu*\bar{y})$. In particular $h(\nu) \neq 0$ and defining $\chi \in S^\wedge$ by $\chi(s) := \dfrac{h(\nu*\bar{s})}{h(\nu)}$ we get $\chi(x) \neq \chi(y)$ and item (iii) follows.

Finally we assume item (iii) and prove item (i). Let μ be any non-zero measure in $M_a(S)$ and $H(S^\wedge)$ be the linear span of $S^\wedge$ in $C(S)$. Then $H(S^\wedge)$ separates points of S and contains the constant semicharacter 1. By Appendix A.4, we have that $H(S^\wedge)$ is (norm) dense in $L^1(S,|\mu|)$. Since $f \to \int f d\mu$ defines a non-zero functional on $L^1(S,|\mu|)$, there exists $g \in H(S^\wedge)$ such that $\int g d\mu \neq 0$. Hence there exists $\chi \in S$ with $\int \chi d\mu \neq 0$. Recalling Theorem 5.3 we have $h(\mu) \neq 0$ for some $h \in M_a(S)^\wedge$. Hence (iii) implies (i).

6. NOTES ON REFERENCES

For locally compact topological semigroups the object $M_a(S)$ was first introduced by A.C. and J.W. Baker ([5] and [6]) and subsequently studied extensively by G.L.G. Sleijpen ([92], [93] and [94]). In particular, for such semigroups S, the results of sections 1 and 2 can be found in [5], [6] and [92]. The extension to include topological semigroups that are not necessarily locally compact is due to Dzinotyiweyi [30].

The object $M_e(S)$ was first introduced by Dzinotyiweyi [32] and was subsequently studied by Dzinotyiweyi and Sleijpen [39]. The results of Section 3 are taken from [32] and [39] while those of Section 4 are taken from [32] and are largely inspired by the results of [92] and [93].

For locally compact topological semigroups S the results of Section 5 can be found in [5], [6] and [92]. The extension to all C-distinguished topological semigroups is taken from [30].

That the spaces $M_a^{\ell}(S)$, $M_e^{\ell}(S)$, $M_a(S)$ and $M_e(S)$ can be identified with $L^1(S)$ in the case of a locally compact group has been known for sometime (see e.g. Hewitt and Ross [59] and Glicksberg [45]).

3 Foundation semigroups and their generalization

1. INTRODUCTION

1.1 DEFINITION. Let S be a topological semigroup with identity element 1 and such that for each neighbourhood U of 1

(a) $x \in \text{int}(U^{-1}(Ux) \cap (xU)U^{-1})$ for all $x \in S$, and

(b) $1 \in \text{int}(U^{-1}v \cap wU^{-1})$ for all $v,w \in U$.

If S is locally compact then S is called a *stip*.

The notion of a stip was introduced by Sleijpen (see e.g. [92] and [94]); its importance stems from the fact that every locally compact foundation semigroup with identity element is a stip (see Theorem 1.2), and a very large class of stips are foundation semigroups (see e.g. Section 2). Further, it still remains an open question whether every stip is a foundation semigroup. (Of course every locally compact topological group G is a stip and one can extract various subgroups of G that are stips.)

This chapter is devoted mainly to a study of the structure and properties of stips. In contrast to Chapter 2, most techniques employed in this Chapter have a large topological (semigroup theoretical) flavour and measure theoretic notions are totally absent in various parts and suppressed where they appear. The results of this section and sections 2 and 3 are taken from work of Sleijpen ([92], [93], [94], and [98]) and those of section 4 are taken from Dzinotyiweyi ([33] and [34]).

1.2 THEOREM. *Let* S *be any topological semigroup with identity element* 1 *and suppose* S *is the foundation of* $M_a(S)$. *Then* S *has properties* (a) *and* (b) *of item* 1.1. *In particular if* S *is also locally compact then* S *is a stip.*

Proof. Let U be a given neighbourhood of 1. By Theorem 2.1.7 we can find a positive measure $\mu \in M_a(S)$ with $D := \text{supp}(\mu)$ compact and $D \subseteq U$. Now item (a) follows trivially from Lemma 2.1.9 (and its right handed analogue).

To prove item (b) we first choose a positive measure $\nu \in M_a(S)$ with $C := \mathrm{supp}(\nu)$ compact and $CC \subset U$. Then mimicing the argument used in the first paragraph of the proof of Theorem 2.1.12, we can find a $y_o \in \mathrm{supp}(\nu * \nu) \subseteq CC$ such that $\nu(y_o C^{-1}) > 0$. So we can find a compact $D \subset y_o C^{-1} \cap C$ such that $\nu(D) > 0$. Hence DD^{-1} is a neighbourhood of 1 by Lemma 2.1.9. But

$$DD^{-1} \subseteq (y_o C^{-1})C^{-1} = y_o(CC)^{-1} \subseteq y_o U^{-1}.$$

Hence $y_o U^{-1}$ is a neighbourhood of 1. It is now trivial to note that item (b) holds.

1.3 Blanket Assumption. *For the remainder of this chapter* S *denotes a stip with identity element* 1 *except where the contrary is explicitly stated.*

2. THE STRUCTURE OF STIPS

We start by introducing and describing a subset S_1 of S that plays a pivotal role in the topological theory of stips.

2.1 NOTATION. We denote by S_1 the set of all $x \in S$ such that $1 \in \mathrm{int}(X^{-1}x \cap xX^{-1})$ whenever X is a neighbourhood of x. (Here the subscript 1 in notation S_1 refers to the identity 1 of S.)

2.2 THEOREM. S_1 *is a dense ideal of* S.

Proof. First we show that $xy \in S_1$ given that $x \in S_1$ and $y \in S$. Let W be a neighbourhood of xy and X a neighbourhood x with $Xy \subseteq W$. By definition of S_1 we then get

$$1 \in \mathrm{int}(xX^{-1}) \subseteq \mathrm{int}((xy)(Xy)^{-1} \subseteq \mathrm{int}((xy)W^{-1}). \qquad (1)$$

Choose neighbourhoods V of x and U of 1 such that $VU \subset X$. Since $V^{-1}x$ and U are neighbourhoods of 1, item 1.1(a) implies that $U^{-1}((V^{-1}x)y)$ is a neighbourhood of y, and so

$$1 \in \mathrm{int}(y^{-1}(U^{-1}((V^{-1}x)y))) \subseteq \mathrm{int}((VUy)^{-1}(xy)) \subseteq \mathrm{int}(W^{-1}(xy)). \qquad (2)$$

From (1) and (2) we conclude that $xy \in S_1$ and so $S_1 S \subseteq S_1$. Similarly $SS_1 \subseteq S_1$.

Second, we show that S_1 is dense in S. It is sufficient to show that $1 \in \bar{S}_1$.

Let U and V be compact neighbourhoods of 1 with $V^2 \subseteq U$. By item 1.1(b) we can find $v, w \in V$ such that $1 \in int(V^{-1}v \cap wV^{-1})$. Then $u := wv$ is in U and

$$1 \in int(V^{-1}v \cap wV^{-1}) \subseteq int((wV)^{-1}(wv) \cap (wv)(Vv)^{-1})$$

$$\subseteq int(U^{-1}u \cap uU^{-1}). \qquad (3)$$

To show that $u \in S_1$, let X be a neighbourhood of u and suppose, on the contrary, $1 \notin int(X^{-1}u)$. Then we can find a net(x_α) such that

$$x_\alpha \to 1 \text{ and eventually } x_\alpha \notin X^{-1}u. \qquad (4)$$

By item (3) we choose a net $(u_\alpha) \subseteq U$ such that

$$\text{eventually } u_\alpha x_\alpha = u.$$

Since U is compact and $x_\alpha \to 1$, we may assume that $u_\alpha \to u$. Thus eventually $u = u_\alpha x_\alpha \in Xx_\alpha$ or $x_\alpha \in X^{-1}u$ (since X is a neighbourhood of u), which contradicts item (4). By this conflict we must have $1 \in int(X^{-1}u)$ and similarly $1 \in int(uX^{-1})$. Hence $u \in S_1 \cap U$ and $1 \in \bar{S}_1$.

We will need the following Proposition in Chapter 5.

2.3 PROPOSITION. *If* J *is a dense (right) ideal of* S *then the interior of* $(S\backslash J)(S\backslash J)^{-1}$ *is empty. In particular* $int((S\backslash S_1)(S\backslash S_1)^{-1}) = \phi$.

Proof. Suppose we had $x \in int((S\backslash J)(S\backslash J)^{-1}) \cap J$. Then, since J is a right ideal, $x(S\backslash J) \subseteq J$ and so $x(S\backslash J) \cap (S\backslash J) = \phi$ or $x \in (S\backslash J)(S\backslash J)^{-1}$. Conflict!

The second part of our Proposition now follows from Theorem 2.2.

The next Theorem says that within the category of dense left or right ideals of S we have that S_1 is the "smallest" such ideal.

2.4 THEOREM. *We have*

$$S_1 = \{x \in S : 1 \in int(S^{-1}x \cup xS^{-1})\}$$

$$= \cap\{J \subseteq S : SJ \subseteq J \text{ and } \bar{J} = S\}$$

$$= \cap\{J \subseteq S : JS \subseteq J \text{ and } \bar{J} = S\}$$

$= \cap\{J \subseteq S : SJS \subseteq J \textit{ and } \bar{J} = S\}.$

Proof. Let $A := \{x \in S : 1 \in \text{int}(S^{-1}x)\}$ and $B := \cap\{J \subseteq S : SJ \subseteq J$ and $\bar{J} = S\}$. We now show that $A = B$.

Let $x \in A$ and $J \subseteq S$ with $\bar{J} = S$ and $SJ \subseteq J$. Then $J \cap (S^{-1}x) \neq \phi$ and so $x \in SJ \subseteq J$. Hence $x \in B$ and $A \subseteq B$.

On the other hand let $x \in B$ and set $I := S\backslash S^{-1}x$. Clearly $SI \subseteq I$ and $x \in I$. So $\bar{I} \neq S$ (by definition of B) and in particular $1 \notin \bar{I}$. Hence

$$1 \in \text{int}(S\backslash\bar{I}) \subseteq S^{-1}x \quad \text{and} \quad x \in A.$$

Thus $B \subseteq A$ and we conclude that $A = B$.

By definition of S_1 and Theorem 2.2 we have that

$$S_1 \subseteq B = A \subseteq S_1.$$

It is now trivial to note that the theorem follows.

As a consequence we have the following factorization of S_1.

2.5 COROLLARY. *We have that* $S_1 = S_1S_1$.

Proof. Clearly $S_1S_1 \subseteq S_1$. Now $\overline{S_1S_1} = S$ and $S(S_1S_1)S \subseteq S_1S_1$ by Theorem 2.2, and hence $S_1 \subseteq S_1S_1$ by Theorem 2.4.

In the situation of a topological group G open sets are well behaved; for instance for every open $U \subset G$ and $x \in G$ we have that xU, $x^{-1}U$, xU^{-1} and $U^{-1}xU$ are open subsets. For stips this can be totally disobeyed even when attention is restricted to compact commutative stips. For an example consider $S := [0,1]$ with the interval topology and multiplication given by $xy := \min\{x+y,1\}$. One can easily note that S is in fact a compact foundation semigroup with identity element. Now $S^{-1}0 = \{0\}$ and $S1 = \{1\}$. With these remarks in mind the following result of Sleijpen [77] is of great interest. Throughout the proof of the Theorem we ask the reader to keep in mind the fact that S_1 is a dense ideal of S (see Theorem 2.2).

2.6 THEOREM. *Let* O, U *be open subsets of* S *and* $x \in S$. *Then*

(i) $x \in int((x(U \cap S_1))u^{-1})$ *for all* $u \in U \cap S_1$,

(ii) $x \in int((xu)(U \cap S_1)^{-1})$ *for all* $u \in U \cap S_1$,

(iii) $x(O \cap S_1)^{-1}$ *is open,*

(iv) $(xO)U^{-1}$ *is open.*

Proof. First we prove items (i) and (ii). Let $u \in U \cap S_1$ and choose $u_1, u_2 \in S_1$ such that $u = u_1u_2$, by Corollary 2.5. Let U_1, U_2, V be open subsets of S such that $u_1 \in U_1$, $u_2 \in U_2$, $1 \in V$ and $VU_1U_2 \subseteq U$. We then have

$$1 \in int(u_1U_1^{-1}) \subseteq int(u(U_1u_2)^{-1}) \subseteq int(u(U_1U_2 \cap S_1)^{-1}),$$

and so by item 1.1(a) we have that

$$\begin{aligned} x \in int(x(u(U_1U_2 \cap S_1)^{-1})V^{-1}) &\subseteq int((xu)(VU_1U_2 \cap S_1)^{-1}) \\ &\subseteq int((xu)(U \cap S_1)^{-1}) \end{aligned}$$

and

$$\begin{aligned} x \in int((xV)(u(U_1U_2 \cap S_1)^{-1})^{-1} &\subseteq int((x(VU_1U_2 \cap VS_1))u^{-1}) \\ &\subseteq int((x(U \cap S_1))u^{-1}). \end{aligned}$$

This completes our proof for (i) and (ii).

Second we prove item (iii). Suppose $t \in x(O \cap S_1)^{-1}$ and choose $w \in O \cap S_1$ with $tw = x$. Let V be a neighbourhood of 1 and W of w with $VW \subseteq O$. By (ii) and the fact that S_1 is dense we can find

$$v \in w(W \cap S_1)^{-1} \cap V \cap S_1.$$

So we can find $y \in W \cap S_1$ with $vy = w$.

We note that $(tv)V^{-1} \subseteq x(O \cap S_1)^{-1}$ for if $p \in (tv)V^{-1}$ then

$$x = tw = tvy \in pVy \subseteq pV(W \cap S_1) \subseteq p(O \cap S_1)$$

and so $p \in x(O \cap S_1)^{-1}$. Hence, with item (ii) in mind, we have that

$$t \in int((tv)V^{-1}) \subseteq int(x(O \cap S_1)^{-1})$$

and so (iii) follows.

Finally we prove item (iv). Let $d \in (xO)U^{-1}$ and choose $a \in U$ and $b \in O$ with $da = xb$. Since S_1 is dense and O and U are open, we can find $c \in S_1$ such that $ac \in U$ and $bc \in O$. Now $d(ac) = x(bc)$ and $ac \in S_1$. Hence $d \in (xO)(U \cap S_1)^{-1}$ and $(xO)U^{-1} \subseteq (xO)(U \cap S_1)^{-1}$ or $(xO)U^{-1} = (xO)(U \cap S_1)^{-1}$. Recalling item (iii) we are done.

2.7 EXAMPLES. We mention some examples of stips, (in fact of foundation semigroups with identity element,) from which the reader can easily note that in general we may have the following situations: $1 \notin S_1$ and so $S_1 \neq S$ even when S is cancellative; S is "far" from being cancellative; $\{1\} \cap E_S$ is not necessarily a G_δ-subset of E_S, (recall that E_S is the set of all idempotents in S with the restriction topology of S); we have compact sets $C, D \subseteq S$ and $x \in S$ such that $C^{-1}D$ and $x^{-1}D$ are not compact; S_1 is not necessarily open or even a G_δ-subset of S.

(i) $S := [0,\infty)$ with the additive operation and the usual topology.

(ii) Let T be any discrete space with zero multiplication $xy := z$ for some fixed $z \in T$. Take S to be T with an adjoined isolated identity 1.

(iii) Let $X_i := \{0,1\}$ with maximum operation and discrete topology. Take $S := \prod_{i \in \Lambda} X_i$ the product semigroup with Λ countable.

(iv) Take S to be as in (iii) but with Λ uncountable.

For many more examples of stips we refer the interested reader to Sleijpen [92].

The following exercises can easily be solved now by routine arguments.

2.8 EXERCISES

(i) For every $e \in E_S$ we have that eSe is a stip with identity element e and $(eSe)_e = eS_1e = eSe \cap S_1$. [Note that $(eSe)_e$ is to eSe what S_1 is to S.]

(ii) If S and T are stips then $S \times T$ is a stip.

(iii) If S is a stip and T is a locally compact quotient semigroup of S with identity element, then T is a stip.

(iv) Let $G_1 := \{x \in S : \text{there exists } y \in S \text{ with } xy = yx = 1\}$. Then G_1 is an algebraic subgroup of S and $G_1 = S^{-1}1 \cap 1S^{-1}$.

If $1 \in S_1$, then G_1 is an open subgroup of S.

If $1 \in S_1$ and S is connected, then S is a group. (See [92] and [94].)

3. THE ROLE OF IDEMPOTENTS IN A STIP

This section is essentially a continuation of the preceding section except that here we focus our studies on the contributions of E_S to the structure of S.

We have already seen that S_1 is a very powerful organ of S (see e.g. Theorem 2.6). If in addition 1 has a countable neighbourhood basis in S a lot more can be said. But then there is a large class of stips without the latter property (see e.g. Example 2.7(iv)). Fortunately most of the studies one can do on a stip with its identity having a countable neighbourhood basis can be extended to a general stip S by use of certain elements of E_S - the so called δ-isolated idempotents. It is our intention in this section to lay the foundation for such an extended theory.

3.1 DEFINITION. An element e in E_S is said to be *δ-isolated* if {e} is a G_δ-set in $eSe \cap E_S$, where $eSe \cap E_S$ is endowed with the restriction topology. We denote the set of all δ-isolated idempotents in S by E_S^δ.

NOTE. The identity element of a locally compact topological group and the identity element 1 of a stip which has a countable neighbourhood base of 1, are δ-isolated idempotents. The identity element of the stip mentioned in Example 2.7(iv) is not δ-isolated.

We now characterize the δ-isolated idempotents partly in terms of S_1 (see Theorem 3.6). Towards this end we first prove some lemmas whose full generality will mainly be needed in later parts of this book.

3.2 LEMMA. *Let* K *be a compact subset of* S *and* $x \in S_1$. *Then* $\cap\{S^{-1}y \cap yS^{-1} : y \in Kx\}$ *is a neighbourhood of* 1.

Proof. By Corollary 2.5 we can write $x = x_1x_2$ for some $x_1, x_2 \in S_1$. From Theorem 2.6 ((iii) and (iv)) we have that $(yx_2S)(xS_1^{-1})^{-1}$ is an open neighbourhood of y, for all $y \in K$.

But

$$(yx_2S)(xS_1^{-1})^{-1} \subseteq (yx_2S)x^{-1},$$

and so the compactness of K implies the existence of a finite set $\{y_1,..,y_n\}$ in K such that

$$K \subseteq \bigcup_{i=1}^{n} (y_i x_2 S)x^{-1}.$$

A routine exercise on definitions now leads to

$$0 \subseteq \cap\{(S^{-1}y \cap yS^{-1} : y \in Kx\} \text{ where } 0 := \bigcap_{i=1}^{n} ((y_i x_2)S_1^{-1}) \cap S_1^{-1}x.$$

Since $y_i x_2 \in S_1$ for $i = 1,...,n$; Theorem 2.6(iii) (together with its right handed analogue) says that 0 is a neighbourhood of 1 and our lemma follows.

3.3 LEMMA. *Let* B *be a* G_δ*-subset of* S *containing* 1. *Then there exist a* δ*-isolated idempotent* e *in* B *and a compact subgroup* G *with identity element* e *such that* $G \subseteq B$ *and* G *is a* G_δ*-set in* eSe.

Proof. There is a sequence $\{V_n\}$ of open subsets of S such that $B = \bigcap_{n=1}^{\infty} V_n$. Inductively we can choose a sequence $\{U_n\}$ of open relatively compact neighbourhoods of 1 such that

$$\bar{U}_{n+1}U_{n+1} \subseteq U_n \cap V_n \quad (n \in \mathbb{N}).$$

Then one can easily see that $H := \bigcap_{n=1}^{\infty} U_n$ is a compact subsemigroup of S contained in B.

Now taking e to be an idempotent in the minimal ideal of H we have that G := eHe is a compact subgroup of S with identity element e (see e.g. Theorem 1.2.3). Since G := eHe = H $\cap$ eSe, H is a G_δ-set in eSe. Since $\{e\} = H \cap (E_S \cap eSe) = eHe \cap E_S$ we have $e \in E_S^\delta$.

3.4 LEMMA. *Let* V *be an open neighbourhood of* 1, M *the closure of a* σ*-compact subset of* S, *and* A *a countable subset of* S_1. *Then there exists a* δ*-isolated idempotent* e *in* V *such that* $MA \subseteq eSe$.

Proof. Let $\{K_n\}$ be a sequence of compact sets such that M is the closure of $\bigcup_{n=1}^{\infty} K_n$. Since $A \subseteq S_1$, Lemma 3.2 implies that we can find a G_δ -subset B such that

$$1 \in B \subseteq V \cap (\cap\{S^{-1}y \cap yS^{-1} : y \in (\bigcup_{n=1}^{\infty} K_n)A\}). \tag{*}$$

By Lemma 3.3 we can find $e \in B \cap E_S^\delta$. It is now easy to note that (*) implies that

$$(\bigcup_{n=1}^{\infty} K_n)A \subseteq eSe,$$

which by the closedness of eSe implies that $MA \subseteq eSe$.

3.5 DEFINITION. A subset F of S is said to be *countably closed* if F contains the cluster points of countable subsets of F.

The *countable closure* of F, denoted by χ_o-clo(F), is thus given by

$$\chi_o\text{-clo}(F) := \{x \in S : x \in \bar{A} \text{ where A is a countable subset of F}\}.$$

Now we observe that for any subsets B and C of S, we have that

$$\chi_o\text{-clo}(B)\chi_o\text{-clo}(C) \subseteq \chi_o\text{-clo}(BC).$$

In particular it follows that the countable closure of an ideal is an ideal. (For instance χ_o-clo(S_1) is an ideal of S.)

3.6 THEOREM. *Let* e *be an idempotent element in* S. *Then the following items are equivalent:*

(i) e *is* δ*-isolated;*

(ii) $e \in \chi_o$-*clo*(S_1);

(iii) *there exists a compact subgroup* G *of* S *such that* e *is the identity element of* G *and* G *is a* G_δ*-set in* eSe.

Proof. That item (iii) implies item (i) follows easily from our definition of a δ-isolated idempotent. Now applying Lemma 3.3 to the stip eSe we have that item (i) implies item (iii). Thus (i) and (iii) are equivalent.

Next we assume item (iii) and prove (ii). Let $\{V_n\}$ be a sequence of open relatively compact sets in eSe such that $G = \bigcap_{n=1}^{\infty} V_n$ and $\bar{V}_{n+1} \subseteq V_n$ for all $n \in \mathbb{N}$. Since $(eSe)_e = eSe \cap S_1$, (recall Exercise 2.8(i)), there exists a sequence $\{h_n\}$ in S_1 such that

$$h_n \in V_n \cap (eSe)_e = V_n \cap eSe \cap S_1$$

for all $n \in \mathbb{N}$. Let $h \in \bigcap_{m \in \mathbb{N}} \text{clo}\{h_n : n \geq m\}$. Then $h \in \bigcap_{m \in \mathbb{N}} \bar{V}_m = G$ and $e \in \text{clo}\{h^k : k \in \mathbb{N}\}$ since G is a compact group. This leads to

$$e \in \text{clo}\{h^k : k \in \mathbb{N}\} \subseteq \text{clo}\{h_n^k : n,k \in \mathbb{N}\} \subseteq \chi_o\text{-clo}(S_1).$$

Thus item (ii) holds.

Finally we assume (ii) and prove (i). Clearly item (ii) implies that $e \in \chi_o\text{-clo}(eS_1e)$ and hence $e \in \chi_o\text{-clo}((eSe)_e)$, by Exercise 2.8(i). So we can find a countable set $A \subseteq (eSe)_e$ with $e \in \bar{A}$. Applying Lemma 3.4 to the stip eSe with $M := \{e\}$ we find a δ-isolated idempotent f in eSe such that $A \subseteq fSf$. Since fSf is closed we have that $e \in fSf$. Consequently $e = f$ and item (i) follows.

3.7 EXERCISES

(i) $\chi_o\text{-clo}(S_1) = \cup\{eSe : e \in E_S^\delta\}$.

(ii) If $e \in E_S^\delta$ then $(eS \cup Se) \cap E_S \subseteq E_S^\delta$.

(iii) If 1 has a countable neighbourhood base then all idempotents in S are δ-isolated.

(iv) Let S be the stip mentioned in Example 2.7(iv). Then

$$S_1 = \{f \in \{0,1\}^\Lambda : f(x) = 0 \text{ for finitely many } x \in \Lambda\},$$

$$\chi_o\text{-clo}(S_1) = \{f \in \{0,1\}^\Lambda : f(x) = 0 \text{ for countably many } x \in \Lambda\}$$

and hence

$$S \neq \chi_o\text{-clo}(S_1).$$

(c.f. [92].)

We conclude this section by giving a proposition whose full generality is mainly needed in Theorem 5.2.8 and is given here to avoid repetition. For our purpose in this chapter we are interested in the first item of the proposition which says that for many stips compact neighbourhoods can be realized as closures of products of compact subgroups and countable subsets. This observation has significant applications in our studies - see e.g. proof of Theorem 4.6.

3.8 PROPOSITION. *Let* S *have a* δ*-isolated identity element* 1, F *be a* G_δ*-subset with* $1 \in F$ *and* A *be a relatively compact subset of* S. *Then there exists a* G_δ *compact subgroup* G *with* $1 \in G$, $G \subseteq F$ *and a countable subset* B *of* A *such that*

(i) $\bar{A} = \overline{GB}$;

(ii) $Gx = xG$ *for all* $x \in \bar{A}$;

(iii) *if* π *is the normalized Haar measure on* G *we have that* $\{f \in C(G\bar{A}) : \bar{x}*\pi(f) = f(x)$ *for all* x *in* $G\bar{A}\}$ *is a (norm) separable subset of* $C(G\bar{A})$.

Proof. Our hypothesis implies the existence of a decreasing sequence $\{V_n\}$ of open relatively compact subsets of S such that

$$\bigcap_{n=1}^{\infty} V_n \subseteq F \quad \text{and} \quad E_S \cap \bigcap_{n=1}^{\infty} V_n = \{1\}.$$

Recalling property (a) in our definition of a stip, an easy induction argument yields an increasing sequence $\{B_n\}$ of finite subsets of A and a decreasing sequence $\{W_n\}$ of open relatively compact neighbourhoods of 1 such that, for each $n \in \mathbb{N}$,

$$\bar{W}_{n+1}^2 \subseteq W_n \cap V_n$$

$$\bar{A} \subseteq (B_n W_n)W_n^{-1} \cap W_n^{-1}(W_n B_n)$$

$$\left.\begin{array}{l} \overline{W_1 A W_1} \cap \bar{W}_{n+1}^{-1}(\bar{W}_{n+1}^{-1} \subseteq (bW_n)W_n^{-1} \\ \overline{W_1 A W_1} \cap (bW_{n+1})\bar{W}_{n+1}^{-1} \subseteq W_n^{-1}(W_n b) \end{array}\right\} \quad \text{for all } b \in B_n.$$

We set $G := \bigcap_{n=1}^{\infty} W_n$ and $B := \bigcup_{n=1}^{\infty} B_n$. Then G is a compact subsemigroup of

S contained in F and has exactly one idempotent, namely 1. Recalling Theorem 1.2.3, we have that G is a compact subgroup.

To prove item (i) let $x \in \bar{A}$ be fixed and W be an open neighbourhood of x. Then GW is an open neighbourhood of Gx. An easy calculation shows that, (by using compactness of $\bar{A}$), we can find $n \in \mathbb{N}$ such that

$$\bar{A} \cap W_n^{-1}(W_n x) \subseteq GW.$$

Since $\bar{A} \subseteq W_n^{-1}(W_n B_n)$, we can find $b \in B_n$ with $x \in W_n^{-1}(W_n b)$. Hence $b \in W_n^{-1}(W_n x)$ and so $B \cap GW \neq \phi$. Consequently $GB \cap W \neq \phi$ and we have $\bar{A} \subseteq \overline{GB}$. Since A is relatively compact and $G\bar{A}$ is compact, we then get

$$\overline{GA} = G\bar{A} \subseteq G\overline{GB} \subseteq \overline{GB} \subseteq \overline{GA} .$$

Hence $\bar{A} = G\overline{GB} \supseteq \overline{GB} \supseteq \bar{A}$. Thus $\bar{A} = \overline{GB}$.

Next we prove item (ii). If $b \in B_n$ then

$$Gb \subseteq \overline{W_1 A W_1} \cap Gb \subseteq \overline{W_1 A W_1} \cap \bar{W}_{k+1}^{-1}(\bar{W}_{k+1} b) \subseteq (bW_k)W_k^{-1}$$

for all $k \in \mathbb{N}$ with $k \geq n$. Now one easily notes that

$$Gb \subseteq \bigcap_{k \geq n} (bW_k)W_k^{-1} \subseteq bG \quad \text{and so} \quad Gb \subseteq bG.$$

Similarly $Gb \supseteq bG$ and we have $Gb = bG$ for all $b \in B$.
For any $g \in G$ and $b \in B$, setting $x := gb$, we get

$$g^{-1}(Gx) = Ggb = Gb = bG \quad \text{and so} \quad Gx = gbG = xG.$$

By compactness of G, we conclude that $Gx = xG$, for all $x \in \overline{GB}$ (i.e. for all $x \in \bar{A}$).

Finally, we prove item (iii). Since $B \times \mathbb{N}$ is countable, by Urysohn's lemma we can find a sequence $\{g_n\}$ of continuous functions from $\bar{A}G$ into $[0,1]$ such that, corresponding to each $(b,m) \in B \times \mathbb{N}$ there is an $n \in \mathbb{N}$ for which

$$g_n(b) = 0, \quad g_n(x) = 0 \quad \text{for all} \quad x \in \bar{A}G \setminus W_n^{-1}(W_n b).$$

[Note that $W_n^{-1}(W_n b)$ is open, by Theorem 2.6(iv).] For each $n \in \mathbb{N}$, let

$$f_n(x) := 2^{-n}(\bar{x}*\pi(g_n)) \quad \text{for all} \quad x \in \bar{A}G.$$

Let $\rho(x,y) := \sum_{n=1}^{\infty} |f_n(x)-f_n(y)|$ for all $x,y \in \bar{A}G$, and note that ρ is a continuous semimetric on $G\bar{A}$. Now one can easily show that every $f \in C(G\bar{A})$, such that $\bar{x}*\pi(f) = f(x)$ $(x \in G\bar{A})$ is continuous with respect to the semimetric ρ. Since $G\bar{A}$ is compact, item (iii) now follows.

4. ACTIONS ON NORMED LINEAR SPACES

We will soon define the *action* of S on a normed linear space A. We are interested in establishing cases where we have weak and norm continuity of such actions. Our results provide a unified approach to various function theoretic results involving uniformly continuous functions on S (as will be seen in Chapter 4) and measure theoretic results involving $M_a(S)$ (as will be seen in Chapter 5).

4.1 DEFINITIONS. Let A be a normed linear space.

We say S has a *left action* (or *anti-action*) on A if there is a map $(s,\mu) \to so\mu$ of $S \times A$ into A such that $xyo\mu = xo(yo\mu)$ or $(xyo\mu = yo(xo\mu)$, respectively), $\|so\mu\| \leq \|\mu\|$ and the map $\mu \to so\mu$ of A into A is linear, for all $s,x,y \in S$ and $\mu \in A$. We also assume that $1o\mu = \mu$ for all $\mu \in A$.

If such an action or anti-action exists we then say $\mu \in A$ has a *norm* (or *weakly*) *separable left orbit over* $U \subseteq S$ if there exists a countable subset C of U such that $\{xo\mu : x \in C\}$ is norm (or weakly, respectively) dense in $\{xo\mu : x \in U\}$. Occasionally we shall simply write "separable" in place of "norm separable". [We shall also assume these definitions for general topological semigroups S.]

4.2 OBSERVATION. We shall formulate our results in terms of actions and remark that these results are also true for anti-actions as one can easily note that a left action for the semigroup (S,.) is a left anti-action for the semigroup (H,*) where H = S and $x*y := y.x$ for all $x,y \in S$.

4.3 LEMMA. *Let* S *have a left action on* A, U *be a subset of* S *and* $\mu \in A$. *Then the following items are equivalent.*

(i) η *has a weakly separable left orbit over* U;

(ii) η *has a norm separable left orbit over* U.

Proof. Evidently (ii) implies (i). We now assume that (i) holds and prove (ii). Let C be a countable subset of U such that $\{x\circ\eta : x \in C\}$ is weakly dense in $\{x\circ\eta : x \in U\}$ and let $\overline{co}(T)$ denote the norm-closure of the convex hull of $T := \{x\circ\eta : x \in C\}$. Then $\overline{co}(T)$ is norm separable. Now

$$\{x\circ\eta : x \in U\} \subseteq \text{weak-cl}(\{x\circ\eta : x \in C\})$$

$$\subseteq \text{weak-cl}(\overline{co}(T)) = \overline{co}(T)$$

by the Hahn-Banach Theorem. So the norm separability of $\overline{co}(T)$ implies that of $\{x\circ\eta : x \in U\}$ and we have item (ii) proved.

4.4 THEOREM. *Let* S *have a left action on* A *and* $\eta \in A$ *be such that the set* $\{s \in S : \|s\circ\eta - y\circ\eta\| \leq \varepsilon\}$ *is closed for all* $y \in S$ *and* $\varepsilon > 0$. *Then item* (i) *implies item* (ii) *and item* (ii) *is equivalent to item* (iii), *where*:

(i) η *has a weakly separable left orbit over a neighbourhood* U *of* 1;

(ii) *the map* $s \to s\circ(x\circ\eta)$ *of* S *into* A *is norm continuous, for all* $x \in S_1$;

(iii) $x\circ\eta$ *has a weakly separable left orbit over some neighbourhood* U *of* 1 *for all* $x \in S_1$.

Proof. Suppose item (i) holds. To prove (ii) it is sufficient to show that the map $s \to s\circ(x\circ\eta)$ is continuous at 1, for all $x \in S_1$. For then if any $x \in S_1$ is fixed and $\varepsilon > 0$ given, we have $W := \{s \in S : \|s\circ(x\circ\eta) - x\circ\eta\| < \varepsilon\}$ a neighbourhood of 1. Now if (s_α) is a net converging to s in S, there is $w \in W \cap S_1$ such that $(s_\alpha) \subseteq (sW)w^{-1}$ eventually, by Theorem 2.6(i). So we can find a net $(w_\alpha) \subseteq W$ such that, eventually

$$s_\alpha w = s w_\alpha.$$

Hence, eventually

$$\begin{aligned}\|s_\alpha\circ(x\circ\eta) - s\circ(x\circ\eta)\| &\leq \|s_\alpha\circ(x\circ\eta) - s_\alpha\circ(w\circ x\circ\eta)\| \\ &\quad + \|s\circ(w_\alpha\circ x\circ\eta) - s\circ(x\circ\eta)\| \\ &\leq \|x\circ\eta - w\circ(x\circ\eta)\| \\ &\quad + \|w_\alpha\circ(x\circ\eta) - x\circ\eta\| \\ &< 2\varepsilon,\end{aligned}$$

and item (ii) would follow.

Now suppose (u_α) is a net converging to 1 in S, let $\varepsilon > 0$ be given and $x \in S_1$ be fixed. Then $V := S^{-1}x \cap U$ is a neighbourhood of 1, by the definition of S_1. So item (i) together with Lemma 4.3 imply that there is a sequence $\{s_n\} \subseteq U$ such that

$$V = \bigcup_{n=1}^{\infty} (O(s_n) \cap V) \quad \text{where} \quad O(s_n) := \{y \in S : \|yo\eta - s_n o\eta\| \leq \varepsilon\}.$$

By the Baire category theorem (see e.g. Kelley [65]) we have $\text{int}(O(s_k) \cap V) \neq \phi$ for some positive integer k. For our convenience let $O := \text{int}(O(s_k) \cap V)$. Since S_1 is dense in S and $O \subseteq S^{-1}x$, we can find $w \in O \cap S_1$ and $y \in S$ such that

$$yw = x \tag{1}$$

Recalling Theorem 2.6(i), we have that $(u_\alpha y) \subseteq (yO)w^{-1}$ eventually. So we can find a net $(z_\alpha) \subseteq O$ such that, eventually

$$u_\alpha yw = yz_\alpha. \tag{2}$$

Hence, eventually

$$\begin{aligned} \|u_\alpha o(xo\eta) - (xo\eta)\| &= \|u_\alpha xo\eta - xo\eta\| \\ &= \|u_\alpha ywo\eta - ywo\eta\| && \text{by (1)} \\ &= \|yz_\alpha o\eta - ywo\eta\| && \text{by (2)} \\ &= \|yo(z_\alpha o\eta) - yo(wo\eta)\| \\ &\leq \|z_\alpha o\eta - wo\eta\| \\ &\leq 2\varepsilon && \text{since } (z_\alpha) \subseteq O \text{ and } w \in O. \end{aligned}$$

As remarked before, this completes our deduction of item (ii) from (i).

A standard argument easily shows that (ii) implies (iii) : for if U is any compact neighbourhood of 1, then (ii) implies that $\{so(xo\eta) : s \in U\}$ is a compact subset of the (metric) space A for all $x \in S_1$.

On the other hand suppose item (iii) holds and let $\varepsilon > 0$ be given. It is evident that, since the set $\{s \in S : \|so\eta - yxo\eta\| \leq \varepsilon\}$ is closed, we have the set $\{s \in S : \|so(xo\eta) - yo(xo\eta)\| \leq \varepsilon\}$ closed, for all $y \in S$ and $x \in S_1$.

Hence the map $s \to so y o(x o \eta)$ of S into A is norm continuous ($x,y \in S_1$), since (i) implies (ii). Now $yo(xo\eta) = yxo\eta$ and $S_1S_1 = S_1$, by Corollary 2.5. So item (ii) holds. This completes our proof.

4.5 LEMMA. *Let G be a compact group with a left action on A and $\eta \in A$. Then the following items are equivalent:*

(i) *the map $x \to xo\eta$ of G into A is weakly continuous;*

(ii) *the map $x \to xo\eta$ of G into A is norm continuous.*

Proof. Evidently (ii) implies (i). Now suppose (i) holds and let $M_n^\ell(G,A) := \{\mu \in A : \text{the map } x \to xo\mu \text{ of G into A is norm continuous}\}$. Because $x \to xo\eta$ is weakly continuous, the weak integral $\nu o\eta := \int xo\eta d\nu(x)$ exists for each $\nu \in M(G)$. (For matters to do with weak integrals (and vector integration) we refer the reader to Diestel and Uhl [27].) Moreover, since the convex-hull of the weakly compact set $\{xo\eta : x \in G\}$ is weakly relatively compact, we have $\nu o\eta \in A$. The map $\nu \to \nu o\eta$ of M(G) into A is clearly (norm to norm) continuous. Now for each $\nu \in M_a(G)$ we have $x \to \bar{x}*\nu$ a norm continuous map of G into M(G), so we conclude the composite

$$x \to \bar{x}*\nu \to xo\nu o\eta$$

is norm continuous into A. Thus $\nu o\eta \in M_n^\ell(G,A)$ for all $\nu \in M_a(G)$.

Now, by definition of the weak integral, we have

$$h(\nu o\eta) = \int h(xo\eta)d\nu(x) \quad (\nu \in M(G) \text{ and } h \in A^*).$$

Since $x \to h(xo\eta)$ is continuous ($h \in A^*$) and G is compact, we easily see that $\nu \to \nu o\eta$ is continuous from M(G) with the weak* topology to A with the weak topology. Hence, since $M_n^\ell(G,A)$ is (norm and hence) weakly closed and $M_a(G)$ is weak* dense in M(G), we conclude that $\eta = 1o\eta$ is in $M_n^\ell(G,A)$ and our lemma is proved.

4.6 THEOREM. *If S has a left action on A and $\eta \in A$ the following items are equivalent:*

(i) *the map $s \to so\eta$ of S into A is weakly continuous;*

(ii) *the map $s \to so\eta$ of S into A is norm continuous.*

Proof. It is clearly sufficient to prove that (i) implies (ii). So suppose $\eta \in A$ satisfies (i). Let $\varepsilon > 0$ be given and $y \in S$ be fixed. We now show that

$$O(y) := \{s \in S : \|so\eta - yo\eta\| \leq \varepsilon\}$$

is closed. So let $x \in \overline{O(y)}$. Setting

$$O(yo\eta) := \{\nu \in A : \|\nu - yo\eta\| \leq \varepsilon\},$$

we note that the Hahn-Banach theorem implies that

$$xo\eta \in \text{weak-cl}(O(yo\eta)) = \text{norm-cl}(O(yo\eta)) = O(yo\eta).$$

Consequently $\|xo\eta - yo\eta\| \leq \varepsilon$ and $x \in O(y)$. So $O(y) = \overline{O(y)}$ and $O(y)$ is closed.

Next we note that $M_n^{\ell}(S,A) := \{\nu \in A : x \to xo\nu$ is a norm continuous map of S into $A\}$ is norm and so weakly closed in A.

Let A be a compact neighbourhood of 1 and suppose (for the moment) that 1 is δ-isolated. Then (the right handed version of) Proposition 3.8(i) says that

$$A = \overline{BG}$$

for some countable $B \subseteq A$ and compact subgroup G of S. By Lemma 4.5 we have that $\{xo\eta : x \in G\}$ is norm compact and so we can find a countable subset D of G such that $\{xo\eta : x \in D\}$ is norm dense in $\{xo\eta : x \in G\}$.
Now for each u in A we have that

$$\begin{aligned} uo\eta &\in \text{weak-cl}(\{bxo\eta : b \in B \text{ and } x \in G\}) \qquad \text{since } A = \overline{BG} \\ &= \text{weak-cl}(\{bo(xo\eta) : b \in B \text{ and } x \in G\}) \\ &= \text{weak-cl}(\{bo(do\eta) : b \in B \text{ and } d \in D\}) \text{ since } \eta \in M_n^{\ell}(G,A) \\ &= \text{weak-cl}(\{xo\eta : x \in BD\}). \end{aligned}$$

Since BD is countable we thus have that η has a weakly separable left orbit over A.

So by Theorem 4.4 we have $xo\eta \in M_n^{\ell}(S,A)$ for all $x \in S_1$. Since S_1 is dense in S and $M_n^{\ell}(S,A)$ is weakly closed, we thus have

$$\eta \in \text{weak-cl}(\{xo\eta : x \in S_1\}) \subseteq M_n^{\ell}(S,A).$$

Thus our theorem follows for the case where 1 is δ-isolated.

Finally we suppose that 1 is not necessarily δ-isolated. Let U be any fixed compact neighbourhood of 1 and $x \in S_1$. Lemma 3.4 says that we can find a δ-isolated idempotent e in S such that $Ux \subseteq eSe$. Thus eSe is a stip with a δ-isolated identity e; so it follows that $\eta \in M_n^{\ell}(eSe,A)$ and hence that $\{yo\eta : y \in Ux\}$ is a norm compact subset of A. Hence $\{uo(xo\eta) : u \in U\}$ contains a countable norm (and so weakly) dense subset. By Theorem 4.4, we thus have $yo(xo\eta) \in M_n^{\ell}(S,A)$ for all $y \in S_1$. Since x was arbitrarily chosen in S_1 and $S_1 = S_1S_1$ we have $so\eta \in M_n^{\ell}(S,A)$ for all $s \in S_1$. Now $\bar{S}_1 = S$, so item (i) and the weak-closure of $M_n^{\ell}(S,A)$ imply that

$$\eta \in \text{weak-cl}(\{so\eta : s \in S_1\}) \subseteq M_n^{\ell}(S,A).$$

Thus (ii) holds and our proof is complete.

4.7 REMARK. Recalling the proof of Lemma 4.5 and the inner regularity of Radon measures, the reader should now be in a position to prove the following generalization of Lemma 2.1.6(ii) : *Let* S *be any topological semigroup acting on a normed linear space* A *such that the map* $x \to xo\eta$ *of* S *into* A *is weakly continuous for some* $\eta \in A$. *Then for all* $\phi \in A^*$ *we have*

$$\phi(\nu o\eta) = \int \phi(xo\eta)d\nu(x)$$

where $\nu o\eta := \int xo\eta d\nu(x)$, *for all* $\nu \in M(S)$.

4 Algebras of functions

Various results on algebras of functions on a topological semigroup S can be found in the literature. These include results on the so called algebras of weakly almost periodic functions, uniformly continuous functions and other subalgebras of C(S). It is not our intention in this chapter to give a complete study of such function algebras; instead, we extract some of those results whose success is largely due to the existence of absolutely continuous measures or a related property on the underlying semigroup. Thus to a large extent this chapter deals with some applications of our studies in Chapters 2 and 3.

1. UNIFORMLY CONTINUOUS FUNCTIONS

First we collect together some definitions and notations.

1.1 DEFINITIONS. Let S be a topological semigroup.

A function f in C(S) is said to be *left uniformly* (or *left weakly uniformly*) *continuous* if the mapping $x \to {}_x f$ of S into C(S) is norm (or weakly respectively) continuous.

We denote the space of all left uniformly continuous functions on S by LUC(S) and that of left weakly uniformly continuous functions by LWUC(S). Similarly one defines the spaces RUC(S) and RWUC(S) of right uniformly (respectively) continuous functions on S.

We say a function is uniformly continuous on S if it belongs to the space UC(S) := LUC(S) ∩ RUC(S), and weakly uniformly continuous if it belongs to the space WUC(S) := LWUC(S) ∩ RWUC(S).

A function f in C(S) is said to be *weakly almost periodic* if the set $\{{}_x f : x \in S\}$ is relatively weakly compact. We denote the set of all weakly almost periodic functions on S by WAP(S).

Although we have defined weakly almost periodicity of a function in terms of its left translates an equivalent definition in terms of right translates is also acceptable. This follows easily from the following result which is an immediate consequence of Grothendieck's Theorem (see Appendix B.7).

1.2 PROPOSITION. *Let* S *be a topological semigroup and* $f \in C(S)$. *Then the following items are equivalent:*

(i) *the set* $\{{}_x f : x \in S\}$ *is relatively weakly compact;*

(ii) *whenever* $\{x_n\}$ *and* $\{y_m\}$ *are sequences in* S *the intersection of the closures of the sets* $\{f(x_n y_m) : n < m\}$ *and* $\{f(x_n y_m) : n > m\}$ *is non-empty.*

(iii) *the set* $\{f_x : x \in S\}$ *is relatively weakly compact.*

As an application of Theorem 3.4.4 we have the following characterization of uniformly continuous functions in terms of separable orbits. First we note that a topological semigroup S has a natural left anti-action on C(S) give by the map $(s,f) \to {}_s f$ of $S \times C(S)$ into C(S). With this in mind we have

1.3 THEOREM. *Let* S *be a stip and* $f \in C(S)$. *Then item* (i) *implies* (ii) *and* (ii) *and* (iii) *are equivalent, where*

(i) f *has a weakly separable left orbit over a neighbourhood of* 1;

(ii) ${}_x f$ *is left uniformly continuous, for all* $x \in S_1$;

(iii) ${}_x f$ *has a weakly separable left orbit over a neighbourhood of* 1, *for all* $x \in S_1$.

Proof. Let $y \in S$ and $\varepsilon > 0$ be given. Then this Theorem will follow from Theorem 3.4.4 if we can show that the set $\{s \in S : \|{}_s f - {}_y f\|_S \leq \varepsilon\}$ is closed. So suppose (s_α) is a net converging to s in S with $\|{}_{s_\alpha} f - {}_y f\|_S \leq \varepsilon$. Hence $|f(s_\alpha z) - f(yz)| \leq \varepsilon$ for all $z \in S$ and so $|f(sz) - f(yz)| \leq \varepsilon$ for all $z \in S$. Thus

$$\|{}_s f - {}_y f\|_S := \sup\{|f(sz) - f(yz)| : z \in S\} \leq \varepsilon$$

and our result follows.

1.4 EXAMPLE. We give an example of a stip S and a function $f \in C(S)$ with a separable left orbit over a neighbourhood of 1 but is such that f is not left uniformly continuous. We consider the following example of a stip S which is also the foundation of $M_a(S)$.

Let

$$S = \{-\frac{1}{2n} : n \in \mathbb{N}\} \cup \{0\} \cup \{\frac{1}{2n+1} : n \in \mathbb{N}\}$$

with the operation

$$xy = yx = x \quad \text{if} \quad |x| \geq |y|, \text{ for all } x,y \in S;$$

and the topology of S coincides with the restriction of the line topology on $\{-\frac{1}{2n} : n \in \mathbb{N}\} \cup \{0\}$ while its restriction on $\{0\} \cup \{\frac{1}{2n+1} : n \in \mathbb{N}\}$ is discrete. Since for all $x \in S\setminus\{0\}$ we have xS finite, $\bar{x} \in M_a(S)$ by Exercise 2.3.10(a)(i). Thus S is the foundation of $M_a(S)$ with identity element 0. Since S is countable, the function $f \in C(S)$ given by

$$f(x) := \begin{cases} 1 & \text{for } x > 0 \\ 0 & \text{for } x \leq 0, \quad (x \in S) \end{cases}$$

has a separable left orbit over S. But if $x_n := -\frac{1}{2n}$ then $x_n \to 0$ while $\|{}_{x_n}f - {}_0f\|_S \to 1$ and so $f \notin LUC(S)$.

1.5. For every topological semigroup S one easily notes that $WAP(S) \subseteq WUC(S)$. So for a stip S, taking $A := C(S)$ in Theorem 3.4.6, we have

THEOREM. *If* S *is a stip, we have that*

(i) $LUC(S) = LWUC(S)$;

(ii) $WAP(S) \subseteq UC(S)$.

1.6. The preceding results are taken from Dzinotyiweyi ([33] and [34]). When S is a locally compact group, Theorem 1.5(ii) has been proved in a different way by Burckel [13].

We now turn to the situation where the underlying topological semigroup S is assumed to be the foundation of $M_a(S)$. Hence we remind the reader that the term "foundation semigroup" mentioned hereafter is as defined in item 2.3.3.

The results given in the remainder of this section are taken from Dzinotyiweyi and Milnes [38].

For any topological semigroup S *whenever we talk of a function* f *in* m(S) *as having a (weakly) separable left orbit, this will be done in terms of the left anti-action of* S *on* m(S) *given by the mapping* $(s,f) \to {}_sf$ *of* $S \times m(S)$ *into* m(S), *for all* $s \in S$ *and* $f \in m(S)$. *Further,* f *is said to be measurable if* f *is* μ*-measurable, for all* $\mu \in M_a(S)$.

1.7. The following remarks will enable us to simplify the proof of our next lemma.

Let S be a foundation semigroup with identity element 1. Let $\mathcal{B}$ be the space of all bounded linear operators from $M_a(S)$ into LUC(S) with the norm operator topology and note that S has a left action on $\mathcal{B}$ given by the mapping $(s,P) \to soP$ of $S \times \mathcal{B}$ into $\mathcal{B}$. Here

$$soP(\nu) := P(\bar{s}*\nu) \quad \text{for all } \nu \text{ in } M_a(S).$$

Let $\mathcal{B}_a := \{P \in \mathcal{B} : \text{the mapping } s \to soP \text{ of } S \text{ into } \mathcal{B} \text{ is continuous}\}$ and hence note that the integral $\nu oP := \int soPd\nu(s)$ exists for all ν in $M_a(S)$ and $P \in \mathcal{B}_a$ (see e.g. [27]). Thus $\mathcal{B}_a$ is an $M_a(S)$-module under the map $(\nu,P) \to \nu oP$. Given $\varepsilon > 0$ and P in $\mathcal{B}_a$ we have that $W := \|s \in S : \|soP - P\| < \varepsilon\}$ is an open neighbourhood of 1 and hence, for ν in $M_a(S)$ with $\nu(W) = \|\nu\| = 1$, we have

$$\|\nu oP - P\| = \left\|\int (soP-P)d\nu(s)\right\| \leq \int \|soP-P\| d\nu(s) < \varepsilon.$$

Thus the $M_a(S)$-module $\mathcal{B}_a$ has a bounded approximate identity, so by Cohen's factorization theorem we have that

$$\mathcal{B}_a = M_a(S) \text{ o } \mathcal{B}_a$$

where $M_a(S) \text{ o } \mathcal{B}_a := \{\nu oP : \nu \in M_a(S) \text{ and } P \in \mathcal{B}_a\}$. (The reader can find details on Cohen's Factorization Theorem for the case of Banach Modules in e.g. Bonsall and Duncan [11].)

1.8 LEMMA. *Let* S *be a foundation semigroup with identity element* 1 *and* $f \in m(S)$ *a function with a separable left orbit over a neighbourhood of* 1. *If* f *is* $M_a(S)$*-measurable, then* ${}_xf$ *is equal almost everywhere* (i.e. a.e. with respect to every measure in $M_a(S)$) *to a function in* LUC(S), *for all* x *in* S_1.

Proof. For every $\nu \in M_a(S)$ and $M_a(S)$-measurable h in m(S) we recall that the functions νoh and $ho\nu$ on S are given by

$$\nu oh(s) := \int h(ys)d\nu(y) \quad \text{and} \quad ho\nu(s) := \int h(sy)d\nu(y) \quad (s \in S).$$

Now one notes that, for all x and y in S,

$$\|{}_x\nu oh - {}_y\nu oh\|_S \leq \|h\|_S \; \|\nu*\bar{x} - \nu*\bar{y}\|$$

and so νoh is in LUC(S) since the map r_ν of S into M(S) is norm continuous by Theorem 2.2.7.

We now show that $xoF \in \mathcal{B}_a$ for all $x \in S_1$, where $F \in \mathcal{B}$ is given by

$$F(\nu) := \nu of \quad (\nu \in M_a(S)).$$

It is trivial to note that, for all x and y in S, we have

$$\|xoF - yoF\| \leq \|{}_xf - {}_yf\|_S$$

and so it follows that F has a separable left orbit over a neighbourhood of 1. Further, given $\varepsilon > 0$ we note that $O(x) := \{s \in S : \|soF - xoF\| \leq \varepsilon\}$ is closed, for all x in S. For if $\nu \in M_a(S)$, then indeed

$$O_\nu(x) := \{s \in S : \|F(\bar{s}*\nu) - F(\bar{x}*\nu)\|_S \leq \varepsilon \; \|\nu\|\}$$

is closed and clearly

$$O(x) = \cap\{O_\nu(x) : \nu \in M_a(S) \text{ and } \|\nu\| = 1\}.$$

Hence, by Theorem 3.4.4, we have that $xoF \in \mathcal{B}_a$, for all x in S_1.

Next we fix $x \in S_1$. Recalling that $\mathcal{B}_a = M_a(S) \circ \mathcal{B}_a$, we can find $\rho \in M_a(S)$ and $H \in \mathcal{B}_a$ such that

$$xoF = \rho oH.$$

Then for every ν in $M_a(S)$ we have that

$$xoF(\nu) := F(\bar{x}*\nu) = \bar{x}*\nu of = \nu o_x f$$

and

$$(\rho oH)(\nu) := \int (yoH)(\nu)d\rho(y) = \int H(\bar{y}*\nu)d\rho(y) = H(\rho*\nu) = H(\rho)o\nu.$$

[The reader may refer to Diestel and Uhl [27] for details on the vector integral.]

So, noting that $h := H(\rho)$ is in LUC(S), we have

$$\nu o_x f = h o \nu \quad \text{for all } \nu \text{ in } M_a(S).$$

In particular

$$\int ({}_x f(s) - h(s)) d\nu(s) = \nu o_x f(1) - h o \nu(1) = 0 \qquad (\nu \in M_a(S)).$$

By Proposition 1.3.3 we conclude that ${}_x f = h$ almost everywhere.

1.9 LEMMA. *Let* S *be a foundation semigroup with identity element* 1, $f \in m(S)$ *and* $h \in C(S)$ *such that* $f = h$ *almost everywhere and* f *has a separable left orbit over a neighbourhood* U *of* 1. *Then* ${}_x f$ *is in* LUC(S) *for all* x *in* S_1.

Proof. Let $x \in S_1$ be fixed. We now show that ${}_x f = {}_x h$, and the result will follow, by Lemma 1.8.

Suppose on the contrary there exists $s_o \in S$ such that

$${}_x f(s_o) \neq {}_x h(s_o).$$

Let W be a neighbourhood of 1 with $WW \subset U$. By Theorems 3.2.2 and 3.2.6(iii) we can find $x_1 \in xS_1^{-1} \cap W \cap S_1$ and hence $x_2 \in S_1$ with $x_1 x_2 = x$. Since $x_1 \in S_1 \cap W$, the definition of S_1 implies that we can find a neighbourhood V of 1 with $V \subset W^{-1} x_1 \cap W$.

By the left translation invariance of $M_a(S)$, the set

$$N(s) := \{y \in S : {}_s f(y) \neq {}_s h(y)\} \text{ is } M_a(S)\text{-negligible.}$$

The given hypothesis implies existence of a sequence $\{s_n\} \subseteq U$ such that $\{{}_{s_n} f : n \in \mathbb{N}\}$ is dense in $\{{}_s f : s \in U\}$. Now $N := \bigcup_{n=1}^{\infty} N(s_n)$ is $M_a(S)$-negligible and so $H := S \backslash N$ is dense in S, since S is a foundation semigroup. By (the right handed version of) Theorem 3.2.6(i) we have $t_1^{-1}(V x_2 s_o)$ a neighbourhood of $x_2 s_o$, for some $t_1 \in V \cap S_1$. Since H is dense, we can find

$$s \in H \cap t_1^{-1}(V x_2 s_o) .$$

Thus $t_1 s = v_1 x_2 s_o$ for some v_1 in V. Since $V \subset W^{-1} x_1$, we can find w in W with $w v_1 = x_1$. Hence

$$w t_1 s = w v_1 x_2 s_o = x_1 x_2 s_o = x s_o.$$

Setting $t := wt_1$ we have $t \in WW \subset U$ and

$$ts = xs_o.$$

Now

$$_tf(s) := f(ts) = f(xs_o) \neq f(xs_o) = h(ts) = {_th}(s).$$

Since $H \setminus N(t)$ is dense in S, we can find a net $(z_\alpha) \subset H \setminus N(t)$ such that $z_\alpha \to s$. Then

$$_tf(z_\alpha) = {_th}(z_\alpha) \to {_th}(s) \neq {_tf}(s).$$

Hence $_tf \notin C(H)$. However the functions $_{s_n}f$ are continuous on H and so we can find a $\delta > 0$ such that $\| _tf - {_{s_n}f} \|_H \geq \delta$ for all n in $\mathbb{N}$. Hence

$$\| _tf - {_{s_n}f} \|_S \geq \| _tf - {_{s_n}f} \|_H \geq \delta > 0, \quad \text{for all } n \in \mathbb{N}.$$

This inequality contradicts the fact that $\{_{s_n}f : n \in \mathbb{N}\}$ is dense in $\{_sf : s \in U\}$. By this conflict our lemma follows.

We recall that for a stip S and $f \in C(S)$, Theorem 1.3 says that f is "very close" to being left uniformly continuous as soon as f has a weakly separable left orbit over a neighbourhood of 1. Our proof there is entirely independent of measure theory and relies more on topological techniques. If in addition S is a foundation semigroup then a more general result can be achieved via measure theoretic techniques. This is the message of our next Theorem.

1.10 THEOREM. *Let* S *be a foundation semigroup with identity element* 1 *and* $f \in m(S)$ *a function with a weakly left separable orbit over a neighbourhood of* 1. *If* f *is* $M_a(S)$*-measurable then* $_sf$ *is left uniformly continuous, for all* x *in* S_1.

Proof. Lemma 3.4.3 says that f has a separable left orbit over some neighbourhood of 1. Lemmas 1.8 and 1.9 then imply that $_{xy}f = {_y}({_x}f)$ is in LUC(S), for all $x,y \in S_1$. Since $S_1S_1 = S_1$ (see Corollary 3.2.5), we conclude that $_xf \in LUC(S)$, for all $x \in S_1$.

If we visualize Theorem 1.10 as the corresponding measure theoretic generalization of Theorem 1.3, then we may take the following result to be the corresponding measure theoretic generalization of Theorem 3.4.4.

1.11 COROLLARY. *Let* S *be a foundation semigroup with identity element* 1 *and let* S *have a left anti-action on a normed linear space* A. *Let* $\eta \in A$ *have a weakly separable left orbit over a neighbourhood of* 1 *and suppose the functions* $s \to \phi(so\eta)$ *of* S *into* $\mathbb{C}$ *are* $M_a(S)$*-measurable, for all* $\phi \in A^*$. *Then* $s \to so(xo\eta)$ *is a norm continuous mapping of* S *into* A *for all* $x \in S_1$.

Proof. Fix $\phi \in A^*$ and consider the function $f : s \to \phi(so\eta)$ of S into $\mathbb{C}$. By Lemma 3.4.3 we may assume that η has a (norm) separable left orbit over a neighbourhood of 1. Now for all $\psi \in C(S)^*$ and $x,y \in S$ we have that

$$|\psi({}_x f) - \psi({}_y f)| \leq \|\psi\| \, \|{}_x f - {}_y f\|_S \leq \|\psi\| \, \|\phi\| \, \|xo\eta - yo\eta\|$$

and so f has a weakly separable left orbit over a neighbourhood of 1. By Theorem 1.10, we have that ${}_x f \in LUC(S)$, for all x in S_1. In particular ${}_x f \in C(S)$ and so the mapping $s \to so(xo\eta)$ of S into A is weakly continuous, for all $x \in S_1$. Recalling Theorem 3.4.6 our result follows.

1.12. In a general sense a function $f \in m(S)$ is said to be *almost periodic* on the semigroup S if the set $\{{}_x f : x \in S\}$ is relatively norm compact. A well known result of von Neumann says that a Haar measurable almost periodic function on a locally compact topological group is continuous. Now if S is a locally compact topological group and ${}_x f \in C(S)$ for some x in S then $f \in C(S)$, and a function is Haar measurable if and only if it is $M_a(S)$ - measurable. Noting that every almost periodic function on a topological semigroup S has a separable left orbit over S, by Theorem 1.10 we obtain (in an elementary manner) the following generalization of von Neumann's result. (Davis [24] also proved von Neumann's result, but the proof is rather long and technical.)

COROLLARY. *Let* S *be a foundation semigroup with identity element* 1 *and* $f \in m(S)$ *an almost periodic function.*

(i) *If* f *is* $M_a(S)$*-measurable, then* ${}_x f$ *is a continuous almost periodic function.*

(ii) *If* S *is a group and* f *is Haar measurable, then* f *is a continuous almost periodic function.*

1.13 REMARK. Since there are non-continuous characters on some locally compact topological groups and characters are almost periodic functions, the measurability condition cannot be completely dropped in Corollary 1.12 and hence in Theorem 1.10. [See e.g. Hewitt and Ross [59, page 405] for the existence of discontinuous characters on the additive group $\mathbb{R}$.]

1.14. We recall that on a locally compact topological space countable unions of meager sets are meager. Now the reader can easily note that; if in the proof of Lemma 1.9, we replace "$M_a(S)$-negligible" by "meager", use Theorem 1.3 in place of Lemma 1.8 and recall Lemma 3.4.3, we have the following extension of Theorem 1.3. (We leave the details as an exercise for the reader.)

PROPOSITION. *Let* S *be a stip and* $f \in m(S)$ *a function that has a left weakly separable orbit over a neighbourhood of* 1 *and such that the set* $\{s \in S : f(s) \neq h(s)\}$ *is meager, for some* $h \in C(S)$. *Then* ${}_xf \in LUC(S)$, *for all* $x \in S_1$.

2. INVARIANT MEANS

Throughout this section let S *be any topological semigroup with an identity element* 1 *and such that* S *is a* k*-space coinciding with the foundation of* $M_a(S)$ *unless otherwise specified.*

In particular, recalling Theorem 2.1.8 and Corollary 2.2.7 we have that S is C-distinguished and $M_a(S) \subseteq M_n(S)$.

2.1 TERMINOLOGIES. Let A be a closed subspace of $M(S)^*$ such that $\bar{x} \odot h \in A$ for all $x \in S$ and $h \in A$, and $e \in A$ where e is the functional which maps μ to $\mu(S)$ for all $\mu \in M(S)$.

An element m of A^* is *a mean* on A if

$$||m|| = m(e) = 1 \quad \text{and} \quad m(f) \geq 0$$

for every non-negative functional f in A.

A *left invariant mean* (LIM) on A is a mean $m \in A^*$ such that

$m(\bar{x} \circ h) = m(h)$ for all $x \in S$ and $h \in A$.

For any subset H of $M(S)$, let $P(H) := \{\mu \in H : \mu \geq 0$ and $\|\mu\| = 1\}$. Now if A is such that $\mu \circ h \in A$ for all $\mu \in H$ and $h \in A$, a $P(H)$-*left invariant mean* ($P(H)$ - LIM) on A is a mean $m \in A^*$ such that

$m(\mu \circ h) = m(h)$ for all $\mu \in P(H)$ and $h \in A$.

In particular a $P(M(S))$ - LIM on A is called *a topological left invariant mean* (TLIM). (Trivially a TLIM is always a LIM.)

If A is a closed translation invariant subspace of $m(S)$ containing all constant functions, then we naturally take a LIM (or (TLIM) on A to be a mean m in A^* such that $m({}_x f) = m(f)$ (or $m(\mu \circ f) = m(f)$, respectively) for all $x \in S$, $f \in A$ and $\mu \in P(M(S))$.

Let H be any locally compact topological semigroup. *Amenability* of H has been defined to mean the existence of a LIM on (a) LUC(H) by e.g. Namioka [74] and Jenkins [63]; (b) MB(H) - the measurable functions in m(H) where H is a subsemigroup of a group, by e.g. Lau [71]; and (c) $M(H)^*$ by e.g. Wong [104].

In the case where H is a group, Greenleaf [53] showed that all such definitions are equivalent and also are equivalent to related definitions for a TLIM. We extend Greenleaf's result to topological semigroups in Theorem 2.4. Our Theorem could also be deduced from Theorems 6.6 and 6.7 of Paterson [80] if S is assumed to be locally compact. The actual presentation given here is taken from Dzinotyiweyi [31].

For a comprehensive account on the subject of invariant means on locally compact topological groups we refer the reader to Greenleaf [53].

2.2 LEMMA. *Let* $f \in A$ *and* $\mu \in M(S)$, *where* A *is any of the spaces* UC(S), LUC(S), RUC(S), C(S) *and* MB(S) := {g ∈ m(S) : g is a Borel function}. *Then*

(i) $\mu \circ f \in A$;

(ii) *if* $\mu, \nu \in M_a(S)$ *and* $h \in M(S)^*$, *then* $\mu \circ h \in LUC(S)$ *and* $\mu \circ h \circ \nu \in UC(S)$.

Proof. (i) Note that for all $x,y \in S$ we have $\mu \circ f_x(y) = (\mu \circ f)_x(y)$ and so $\|(\mu \circ f)_x - (\mu \circ f)_y\| = \|\mu \circ f_x - \mu \circ f_y\|_S \leq \|\mu\| \; \|f_x - f_y\|$.

Hence item (i) holds for A = RUC(S).

Also item (i) holds for A = C(S) by Lemma 1.3.4.

Next we show that $f \in$ LUC(S) implies that $\mu o f \in$ LUC(S). Evidently LUC(S) is norm and hence weakly closed. If on the contrary $\mu o f \notin$ LUC(S), the Hahn-Banach Theorem requires the existence of a $\phi \in$ LUC(S)* such that $\phi(\mu o f) \neq 0$ and $\phi = 0$ on LUC(S). Then recalling Remark 3.4.7 we would get the ridiculous situation

$$0 \neq \phi(\mu o f) = \int \phi(\bar{x} o f) d\mu(x) = 0.$$

By this conflict we conclude that $\mu o f \in$ LUC(S).

Consequently item (i) also follows for A = UC(S), and it remains to show that $f \in$ MB(S) implies $\mu o f \in$ MB(S). To the latter end, we first note that we may assume μ to be a positive measure without losing generality. Now for any closed set $F \subset S$ we have that the function $\mu o \chi_F$ is upper semicontinuous, by Lemma 1.4.1(ii), so $\mu o \chi_F$ is Borel measurable.

Next if $\{F_n\}$ is an increasing (or decreasing) sequence of Borel sets with $\mu o \chi_{F_n}$ measurable ($n \in \mathbb{N}$) then, for all $s \in S$,

$$\mu o \chi_{F_n}(s) := \int \chi_{F_n}(xs) d\mu(x) \to \mu o \chi_F(s),$$

where $F := \bigcup_{n=1}^{\infty} F_n$ (or $F := \bigcap_{n=1}^{\infty} F_n$, respectively). Recalling the monotone class lemma and the fact that closed subsets generate the Borel subsets of S we have $\mu o \chi_B \in$ MB(S) for every Borel set $B \subset S$. Finally if $\{f_n\}$ is a bounded sequence of simple Borel functions converging pointwise to f, then $\mu o f_n(s) \to \mu o f(s)$ for all $s \in S$ and so $\mu o f \in$ MB(S).

(ii) For all $x,y \in S$ we have

$$\|{}_x(\mu o h) - {}_y(\mu o h)\|_S = \sup\{|h(\mu * \overline{xs}) - h(\mu * \overline{ys})| : s \in S\}$$
$$\leq \|h\| \; \|\mu * \bar{x} - \mu * \bar{y}\|.$$

Since $M_a(S) \subseteq M_n(S)$, we have $\mu o h \in$ LUC(S). Noting that

$$\mu o h o \nu = (\mu \odot h) o \nu = \mu o (h \odot \nu)$$

and $\mu \odot h$, $h \odot \nu \in M(S)^*$, it follows that $\mu o h o \nu \in$ UC(S).

The argument involving MB(S) in our Lemma is taken from Paterson [80].

2.3 LEMMA. *Let* m *be a* LIM *on* UC(S). *Then*

(i) m *is a* TLIM *on* UC(S);

(ii) $m(\mu o h o \nu) = m(\eta o h o \nu)$ *for all* $\nu,\mu,\eta \in P(M_a(S))$ *and* $h \in M(S)^*$.

Proof. For any $\mu \in P(M(S))$ and $f \in UC(S)$ we have

$$m(\mu o f) = \int m(\bar{x} o f) d\mu(x) \quad \text{by Remark 3.4.7}$$
$$= m(f)\mu(S) = m(f);$$

and so item (i) follows.

We now prove item (ii). By item 2.2.11 we may take (ρ_α) to be a bounded (norm) approximate identity for $M_a(S)$. Then

$$|m(\mu o h o \nu) - m(\rho_\alpha * \mu o h o \nu)| \leq \sup\{ \|h\| \ \|\mu * \bar{x} * \nu - \rho_\alpha * \mu * \bar{x} * \nu\| : x \in S\}$$
$$\leq \|h\| \ \|\mu - \rho_\alpha * \mu\| \to 0,$$

(with notation as stated in the Lemma).
By (i), we have

$$m(\rho_\alpha * \mu o h o \nu) = m(\mu o (\rho_\alpha o h o \nu)) = m(\eta o (\rho_\alpha o h o \nu)) = m(\rho_\alpha * \eta o h o \nu)$$

and so

$$|m(\mu o h o \nu) - m(\eta o h o \nu)| \leq |m(\mu o h o \nu) - m(\rho_\alpha * \mu o h o \nu)| + |m(\rho_\alpha * \eta o h o \nu) - m(\eta o h o \nu)|$$
$$\to 0$$

and our result follows.

2.4 THEOREM. *Let* A_1 *and* A_2 *be any of the spaces*: UC(S), LUC(S), RUC(S), C(S), MB(S), $M_a(S)^*$ *and* $M(S)^*$. *Then the following items are equivalent*:

(i) *There exists a* $P(M_a(S))$ - LIM *on* A_1;

(ii) *there exists a* LIM *on* A_2;

(iii) *there exists a* TLIM *on* A_2.

Proof. We note that $UC(S) \subseteq A_1 \subseteq M(S)^*$ for all choices of A_1. So the existence of any of the three types of left invariant mean on $M(S)^*$ implies (by restriction) the existence of the same type of left invariant mean on A_1.

Now let m be a $P(M_a(S))$ - LIM on UC(S). Then, with the above remarks in mind and recalling that a LIM on UC(S) is a TLIM, our Theorem will follow if we can show that there exists a TLIM on M(S)*. To this end we fix measures η and ν in $P(M_a(S))$ and define the function M on M(S)* by

$$M(h) := m(\eta o h o \nu) \quad \text{for all} \quad h \in M(S)^*.$$

Since $\eta o h o \nu \in UC(S)$, we evidently have that M is a mean on M(S)*. Further, for any $\mu \in P(M(S))$ we have that $\mu * \eta \in P(M_a(S))$ and so

$$\begin{aligned} M(\mu o h) := m(\eta o(\mu o h)o \nu) &= m(\mu * \eta o h o \nu) \\ &= m(\eta o h o \nu) \quad \text{by Lemma 2.3(ii)} \\ &= M(h). \end{aligned}$$

This completes our proof.

2.5 SOME OPEN PROBLEMS. Of course the success of Theorem 2.4 is pivoted around the object $M_a(S)$. It is natural to ask what the situation would be when existence of $M_a(S)$ is not assumed. For instance in Chapter 3 we studied an important class of semigroups S which behave as if they were foundation semigroups though the latter is still unknown. The following problem is therefore reasonable.

PROBLEM A. Let S be a stip. We conjecture that

(i) there is a LIM on C(S) if UC(S) has a LIM;

(ii) there is a TLIM on C(S) if there exists a LIM on C(S).

Should the question whether every stip is a foundation semigroup turn out to be answered affirmatively then Problem A would follow. However it would still be interesting if Problem A could be settled without the use of $M_a(S)$.

The preceeding remarks invoke another question. As stated before, our Theorem 2.4 is a generalization of a result well known for groups - see e.g. Greenleaf [53, Theorem 2.2.1]. Now for a topological group G we know that G is locally compact if and only if $M_a(G)$ is non-zero. Hence considering the situation where $M_a(G)$ is zero, we have the following problem.

PROBLEM B. Let G be any (non-locally compact) topological group. We do not know whether the existence of a LIM on C(G) implies the existence of a TLIM on C(G). Further, we do not know whether there is a LIM on C(G) as soon as there is a LIM on UC(G).

2.6 TERMINOLOGY. First we warn the reader that, for the remainder of this section, S is not necessarily as stated at the beginning of the section. Let S be any locally compact topological semigroup and assume the notations introduced in item 2.1 for this S.

A net or sequence $(\mu_\alpha) \subset P(M_a(S))$ is said to be *weakly* (or *strongly*) *convergent to topological invariance* if $\nu*\mu_\alpha - \mu_\alpha \to 0$ and $\mu_\alpha*\nu - \mu_\alpha \to 0$ weakly (or strongly, respectively) in $M_a(S)$ for all $\nu \in M_a(S)$.

Let IM(WUC(S)) denote the set of (left and right) invariant means on WUC(S).

We now commence an argument aimed at estimating the size of IM(WUC(S)).

2.7 LEMMA. *Let S be a locally compact topological semigroup. If S is not relatively neo-compact, then*

$$m(f) = 0 \textit{ for all } f \in C_o(S) \textit{ and } m \in IM(WUC(S)).$$

Proof. Let $f \in C_{oo}(S)$ be positive and put $K := \text{supp}(f)$. Since S is not relatively neo-compact, we can choose a sequence $\{x_n\}$ in S such that

$$x_{n+1} \notin K(x_1^{-1}K \cup \ldots \cup x_n^{-1}K)^{-1} \quad \text{for all } n \in \mathbb{N}.$$

So if $n \neq k$ we have $x_n^{-1}K \cap x_k^{-1}K = \phi$ or, equivalently,

$$\text{supp}({}_{x_n}f) \cap \text{supp}({}_{x_k}f) = \phi.$$

Consequently

$$nm(f) = m({}_{x_1}f + \ldots + {}_{x_n}f) \leq \|f\|_S \quad \text{for all } n \in \mathbb{N},$$

and so $m(f) = 0$.

Since $C_{oo}(S)$ is dense in $C_o(S)$, the remainder of our result follows trivially. (Note that $C_o(S) \subseteq WUC(S)$, by Lemma 1.3.4 and the fact that $C_o(S)* = M(S)$.)

2.8 LEMMA. *Let S be a C-distinguished topological semigroup. There is a net in* $P(M_a(S))$ *weakly convergent to topological invariance if and only if there is a net in* $P(M_a(S))$ *strongly convergent to topological invariance.*

Proof. Suppose $(\eta_\alpha) \subset P(M_a(S))$ converges weakly to topological invariance. Setting $M_{\nu,\mu} := M_a(S)$ we form the locally convex product space $M := \Pi\{M_{\nu,\mu} : (\nu,\mu) \in P(M_a(S)) \times P(M_a(S))\}$ with the product of norm topologies and define the linear map $L : M_a(S) \to M$ by

$$L(\phi)(\nu,\mu) := \nu*\phi*\mu - \phi \text{ for all } \phi \in M_a(S) \text{ and } \nu,\mu \in P(M_a(S)).$$

Appendix B.3 says the weak topology on M coincides with the product of the weak topologies on the $M_{\nu,\mu}$'s. Since $\nu*\eta_\alpha*\mu - \eta_\alpha \to 0$ weakly, for all $\nu,\mu \in P(M_a(S))$, $0 \in$ weak-closure $L(P(M_a(S)))$. Since $L(P(M_a(S)))$ is a convex subset of the locally convex space M, we have weak-closure $L(P(M_a(S))) =$ strong-closure $L(P(M_a(S)))$, by the Hahn-Banach Theorem. So there exists a net $(\rho_\alpha) \subset P(M_a(S))$ such that $L(\rho_\beta) \to 0$ in M or, equivalently, such that

$$\|\nu*\rho_\beta*\mu - \rho_\beta\| \to 0 \quad \text{for all } \nu,\mu \in P(M_a(S)).$$

Hence

$$\begin{aligned}\|\nu*\rho_\beta - \rho_\beta\| &\leq \|\nu*\rho_\beta - \nu*\nu*\rho_\beta*\nu\| + \|(\nu*\nu)*\rho_\beta*\nu - \rho_\beta\| \\ &\leq \|\rho_\beta - \nu*\rho_\beta*\nu\| + \|(\nu*\nu)*\rho_\beta*\nu - \rho_\beta\| \to 0\end{aligned}$$

for all $\nu \in P(M_a(S))$. Similarly $\|\rho_\beta*\nu-\rho_\beta\| \to 0$, and the remainder of our lemma follows trivially.

2.9 LEMMA. *Let S be a locally compact topological semigroup that is not relatively neo-compact and let* $\{\mu_n\}$ *be a sequence in* $P(M_a(S))$ *strongly convergent to topological invariance and such that* $T_n := \text{supp}(\mu_n)$ *is compact, for all* $n \in \mathbb{N}$. *Then given* $n_o \in \mathbb{N}$ *and* $\varepsilon > 0$, *we can find* $n > n_o$ *such that*

$$\mu_n((T_1 \cup \ldots \cup T_{n_o}) \cap T_n) < \varepsilon .$$

Proof. Suppose, on the contrary, there exists an n_o and $\varepsilon > 0$ such that

$$\mu_n((T_1 \cup \ldots \cup T_{n_o}) \cap T_n) \geq \varepsilon \quad \text{for all } n > n_o.$$

Let $f \in C_o(S)$ be a positive function with $f = 1$ on $T_1 \cup \ldots \cup T_{n_o}$ and note that

$$\mu_n(f) \geq \varepsilon \quad \text{for all} \quad n > n_o.$$

Let m be any weak*-cluster point of $\{\mu_n\}$ in WUC(S)* and note that m is an invariant mean on WUC(S) such that $m(f) \geq \varepsilon$. By Lemma 2.7, we must have $m(f) = 0$. This contradiction implies our result.

2.10 LEMMA. *Let S be a σ-compact locally compact topological semigroup (with $M_a(S)$ non-zero) and let there be an invariant mean on* WUC(S). *Then there exists a sequence (ρ_n) in $P(M_a(S))$ converging strongly to topological invariance and such that $K_n := \text{supp}(\rho_n)$ is compact, for all $n \in \mathbb{N}$, if any one of the following conditions holds:*

(a) *S is a foundation semigroup with an identity element.*

(b) *The centre of $F_a(S)$ is not $M_a(S)$-negligible.*

Proof. First we show that there exists a net in $P(M_a(S))$ weakly convergent to topological invariance, if (a) or (b) holds. To this end, suppose (a) holds. Then there exists a topological (left and right) invariant mean m on $M_a(S)^*$, by Theorem 2.4. Now $m \in$ weak*-closure $(P(M_a(S)))$ in $M_a(S)^{**}$. Consequently, there exists a net (μ_α) in $P(M_a(S))$ such that $\mu_\alpha(h) \to m(h)$ for all $h \in M_a(S)^*$. In particular, for each $\nu \in P(M_a(S))$ we have

$$|h(\nu*\mu_\alpha - \mu_\alpha)| = |\mu_\alpha(\nu\circ h) - \mu_\alpha(h)| \to |m(\nu\circ h) - m(h)| = 0,$$

for all $h \in M_a(S)^*$. Similarly $\mu_\alpha*\nu - \mu_\alpha \to 0$ weakly. Thus $(\mu_\alpha) \subset P(M_a(S))$ is weakly convergent to topological invariance.

Next suppose condition (b) holds. Then we can choose $\tau \in P(M_a(S))$ such that $\text{supp}(\tau) \subseteq Z(F_a(S))$, where $Z(F_a(S))$ denotes the centre of $F_a(S)$. We then have $\tau*\nu = \nu*\tau$ for all $\nu \in P(M_a(S))$. Now if m_o is any invariant mean on WUC(S) we have that m_o is topologically invariant, by arguing as in Lemma 2.3(i). Let (η_α) be a net in $P(M_a(S))$ such that $\eta_\alpha(f) \to m_o(f)$, for all $f \in$ WUC(S). Then, for any $\nu \in P(M_a(S))$ and $h \in M_a(S)^*$, we have that $\tau*\nu o h o \tau$, $\tau o h o \tau \in$ WUC(S) by Lemma 3.2; hence, if $\mu_\alpha := \tau*\eta_\alpha*\tau$,

$$\begin{aligned} h(\nu*\mu_\alpha - \mu_\alpha) &= h(\nu*\tau*\eta_\alpha*\tau - \tau*\eta_\alpha*\tau) \\ &= h(\tau*\nu*\eta_\alpha*\tau) - h(\tau*\eta_\alpha*\tau) \\ &= \eta_\alpha(\tau*\nu o h o \tau) - \eta_\alpha(\tau o h o \tau) \\ &\to m_o(\nu o(\tau o h o \tau)) - m_o(\tau o h o \tau) = 0. \end{aligned}$$

Similarly $h(\mu_\alpha*\nu - \mu_\alpha) \to 0$. Thus (μ_α) is weakly convergent to topological invariance.

Now suppose either (a) or (b) holds. Then there exists a net $(\eta_\beta) \subseteq P(M_a(S))$ strongly convergent to topological invariance, by Lemma 2.8. Fix any $\lambda \in P(M_a(S))$ and set $\mu_\beta := \lambda*\eta_\beta*\lambda$. Note that (μ_β) is also strongly convergent to topological invariance. As S is σ-compact, we can choose an increasing sequence of compact neighbourhoos $D_1 \subseteq D_2 \subseteq \ldots$, such that $S = \bigcup_{n=1}^{\infty} D_n$. Noting that the maps $x \to \bar{x}*\mu_\beta$ and $y \to \mu_\beta*\bar{y}$ of S into $M_a(S)$ are norm continuous (see e.g. Corollary 2.2.4) we can choose a sequence $\{\mu_{\beta_1}, \mu_{\beta_2}, \ldots\}$ from the μ_β's such that

$$\|\bar{x}*\mu_{\beta_n}*\bar{y} - \mu_{\beta_n}\| < \frac{1}{5n} \quad \text{for all } x,y \in D_n.$$

Choose compact sets K_n such that

$$\mu_{\beta_n}(S\setminus K_n) < \frac{1}{5n} \text{ for all } n \in \mathbb{N}.$$

Setting $\rho_n := \mu_{\beta_n}(K_n)^{-1}\, \mu_{\beta_n}|K_n$ we have $(\rho_n) \subseteq P(M_a(S))$, $\text{supp}(\rho_n) \subseteq K_n$ and a standard technical argument shows that

$$\|\mu_{\beta_n} - \rho_n\| < 2\mu_{\beta_n}(S\setminus K_n) < \frac{2}{5n}.$$

Consequently

$$\begin{aligned} \|\bar{x}*\rho_n*\bar{y} - \rho_n\| &\le \|\bar{x}*\rho_n*\bar{y} - \bar{x}*\mu_{\beta_n}*\bar{y}\| + \|\bar{x}*\mu_{\beta_n}*\bar{y} - \mu_{\beta_n}\| + \|\mu_{\beta_n} - \rho_n\| \\ &\le 2\|\rho_n - \mu_{\beta n}\| + \|\bar{x}*\mu_{\beta_n}*\bar{y} - \mu_{\beta_n}\| \\ &< \frac{4}{5n} + \frac{1}{5n} = \frac{1}{n} \quad \text{for all } x,y \in D_n. \end{aligned}$$

Now for any $\nu,\eta \in P(M_a(S))$ with compact supports we have $\mathrm{supp}(\nu) \cup \mathrm{supp}(\eta) \subseteq D_n$ for n larger than some n_o and hence

$$\begin{aligned}\|\nu*\rho_n*\eta-\rho_n\| &:= \sup\{|\nu*\rho_n*\eta(f)-\rho_n(f)| : f \in C_o(S), \|f\|_S \leq 1\} \\ &\leq \sup\{\int|\bar{x}*\rho_n*\bar{y}(f)-\rho_n(f)|d\nu(x)d\eta(y): f \in C_o(S), \|f\|_S \leq 1\} \\ &< \frac{1}{n}.\end{aligned}$$

It is now trivial to complete the proof of our lemma.

We are now in a position to prove our second Theorem in this section. But first we mention the following notation needed in the Theorem.

Let $F := \{\phi \in (\ell^\infty)^* : \phi \geq 0, \|\phi\| = 1$ and $\phi(g) = 0$ for all $g \in \ell^\infty$ such that $g(k) \to 0$ as $k \to \infty\}$ and c denote the cardinal of continuum.

2.11 THEOREM. *Let* S *be a* σ*-compact locally compact topological semigroup such that* S *is not relatively neo-compact,* $M_a(S)$ *is non-zero and there exists an invariant mean on* WUC(S). *Suppose either*

(a) S *is a foundation semigroup with identity element, or*

(b) *the centre of* $F_a(S)$ *is not* $M_a(S)$*-negligible.*

Then there exists a linear isometry $\tau : (\ell^\infty) \to \mathrm{WUC}(S)^*$ *such that*

$\tau(F) \subseteq \mathrm{IM}(\mathrm{WUC}(S))$ *and so* $\mathrm{card}(\mathrm{IM}(\mathrm{WUC}(S))) \geq 2^c$.

Proof. Note that we have the hypothesis of Lemma 2.10 met, and let $\{\rho_n\}$ and $\{K_n\}$ be as in Lemma 2.10. Choose $\nu \in P(M_a(S))$ with $C := \mathrm{supp}(\nu)$ compact and note that the sequence $\{\nu*\rho_n*\nu\}$ also converges strongly to topological invariance.

Observing that $T_n := \mathrm{supp}(\nu*\rho_n*\nu) = CK_nC$ is compact $(n \in \mathbb{N})$ and recalling Lemma 2.9, there exists a subsequence $\{n_k\}$ of $\{n\}$ such that

$$\nu*\rho_{n_k}*\nu\,(T_{n_k}\setminus(T_{n_1}\cup\ldots\cup T_{n_{k-1}})) > 1-2^{-k} \quad \text{for } k = 2,3,\ldots .$$

Let $T_{n_o} := \phi$, $F_k := T_{n_k}\setminus(T_{n_1}\cup\ldots\cup T_{n_{k-1}})$ and $\mu_k := \nu*\rho_{n_k}*\nu$, for all $k \in \mathbb{N}$.

Let $\pi : WUC(S) \to \ell^\infty$ be the linear mapping defined by

$$\pi(f)(k) := \rho_{n_k}(f), \quad \text{for all } f \in WUC(S) \text{ and } k \in \mathbb{N}.$$

To see that π is onto, let $g \in \ell^\infty$ be fixed. We now define $\tilde{g} \in \ell^\infty$ as follows: Let

$$\tilde{g}(1) := g(1) \text{ and } \tilde{g}(k) := g(k) - \sum_{i=1}^{k-1} \tilde{g}(i)\mu_k(F_i)/\mu_i(F_i), \text{ for } k = 2,3,\ldots$$

(Indeed $\tilde{g} \in \ell^\infty$ since, by our definition of the F_k's we have $|\tilde{g}(1)| \leq 2\,\|g\|_\infty$ and on assuming $|\tilde{g}(i)| \leq 2\,\|g\|_\infty$ for $i = 1,2,\ldots,k-1$ we are led to

$$|\tilde{g}(k)| \leq \|g\|_\infty + 2\,\|g\|_\infty \sum_{i=1}^{k-1} \mu_k(F_i)/\mu_i(F_i) \leq \|g\|_\infty (1+2 \sum_{i=1}^{k-1} 2^{-k}/(1-2^{-i})) \leq 2\|g\|_\infty.)$$

Now the pairwise disjointness of members of the sequence $\{F_k\}$, that $\tilde{g} \in \ell^\infty$ and $\mu_k(F_k) > \frac{1}{2}$, imply that the function

$$h := \sum_{k=1}^{\infty} (\tilde{g}(k)/\mu_k(F_k))\chi_{F_k}$$

is a linear functional in $M(S)^*$.

Consequently $\nu o h o \nu \in WUC(S)$, by Lemma 3.2. Now, for all $k \in \mathbb{N}$, we have

$$\begin{aligned}\pi(\nu o h o \nu) &= \rho_{n_k}(\nu o h o \nu) \\ &= \nu * \rho_{n_k} * \nu(h) \\ &= \mu_k\Big(\sum_{i=1}^{\infty} (\tilde{g}(i)/\mu_i(F_i))\chi_{F_i}\Big) \\ &= \tilde{g}(k) + \sum_{i=1}^{k-1} \tilde{g}(i)\mu_k(F_i)/\mu_i(F_i) = g(k).\end{aligned}$$

Thus π maps the function $\nu o h o \nu$ onto g and π is onto.

Further, we clearly have

$$\|g\|_\infty = \|\pi(\nu o h o \nu)\|_\infty = \|\nu o h o \nu\|_S.$$

It follows that the dual map $\pi^* : (\ell^\infty)^* \to WUC(S)^*$ is a linear isometry.

To see that $\pi^*\mathcal{F} \subseteq IM(WUC(S))$, let $\phi \in \mathcal{F}$ be fixed. Then, clearly

$$\pi^*\phi \geq 0 \quad \text{and} \quad \pi^*\phi(1) = \phi(1) = 1.$$

Now for any $\eta \in P(M_a(S))$ and $f \in WUC(S)$ we have

$$\pi(\eta o f - f)(k) := \rho_{n_k}(\eta o f - f) = (\eta * \rho_{n_k} - \rho_{n_k})(f) \to 0 \quad \text{as } k \to \infty.$$

Recalling the definition of $\mathcal{F}$ we have

$$\pi^*\phi(\eta o f - f) = 0.$$

Similarly $\pi^*\phi(f o \eta - f) = 0$ and so $\pi^*\phi \in IM(WUC(S))$.

Taking $\beta\mathbb{N}$ to be the Stone-Cech compactification of $\mathbb{N}$, we have $\beta\mathbb{N}\setminus\mathbb{N} \subseteq \mathcal{F}$. Since $card(\beta\mathbb{N}\setminus\mathbb{N}) = 2^c$ and π^* is an isometry, we thus get $card(IM(WUC(S))) \geq card(\mathcal{F}) \geq 2^c$. So $\tau := \pi^*$ is the required map and our Theorem is proved.

2.12 COROLLARY. *Let* S *be a non-relatively neo-compact, σ-compact, locally compact topological semigroup with respect to which the hypothesis of Theorem* 2.11 *holds and let* $E(S) := \{f \in WUC(S) : m(f) = \text{a constant as } m \text{ runs through } IM(WUC(S))\}$. *Then the quotient space* $WUC(S)/_{E(S)}$ *is nonseparable.*

Proof. Suppose, on the contrary, $WUC(S)/_{E(S)}$ is separable. Then there exists a countable set $F := \{f_n : n \in \mathbb{N}\}$ in WUC(S) such that $E(S) + L(F)$ is dense in WUC(S), where $L(F)$ denotes the linear span of F. So each $\rho \in IM(WUC(S))$ is determined by the sequence $\{\rho(f_n)\}$ and so $card(IM(WUC(S))) \leq c$. This contradicts Theorem 2.11.

2.13 NOTES. Various results on the sizes of sets of invariant means can be found in the literature : for discrete semigroups see e.g. Chou ([15] and [16]), Granirer [51] and Klawe [68], and for locally compact topological groups see e.g. Chou ([16] and [17]) and Granirer [49]. Our Theorem 2.11 generalizes some of these results and our techniques follow closely those given in [17]. For gorups, Corollary 2.12 was first proved by Granirer [49]. The proof of our Corollary 2.12 is similar to that mentioned for groups in [18].

3. THE SIZE OF THE DIFFERENCE WUC(S)\WAP(S)

In this section, for a large class of topological semigroups S, we indicate how one can use absolutely continuous measures to explicitly construct many functions in WUC(S)\WAP(S).

For convenience we assume that all the spaces of measures, functions (and functionals) mentioned in this section are real-valued and remark that the generalization to the complex-valued case can be achieved trivially.

First we prove a lemma which plays a pivotal role in the construction of functions in WUC(S)\WAP(S).

3.1 LEMMA. *Let* S *be any topological semigroup which is not relatively neo-compact,* C_o *and* D_o *any fixed compact subsets of* S. *Let* $C := C_o \cup \{1\}$ $D := D_o \cup \{1\}$ *where* 1 *is an identity element of* S *(if there is one) or an adjoined isolated identity element of* S. *Then there exist infinite sequences* $\{x_1, x_2, \ldots\}$ *and* $\{y_1, y_2, \ldots\}$ *in* S *such that*

$$C^{-1}(Cx_n y_m D)D^{-1} \cap C^{-1}(Cx_i y_j D)D^{-1} = \phi$$

if any one of the following three conditions holds:

(a) $n \leq m$ *and* $i > j$;

(b) $n > m$, $i > j$ *and* $n \neq i$;

(c) $n \leq m$, $i \leq j$ *and* $m \neq j$.

Proof. Our proof is by induction. Suppose, by the inductive hypothesis, we have finite sequences $X_p := \{x_1, x_2, \ldots, x_p\}$ and $Y_p := \{y_1, y_2, \ldots, y_p\}$ in S such that the lemma holds for n,m,i,j in $\{1,2,\ldots,p\}$. For convenience, let

$$L_p : \bigcup_{n=1}^{p} \bigcup_{n \leq m \leq p} Cx_n y_m D \quad \text{and} \quad R_p := \bigcup_{m=1}^{p-1} \bigcup_{p \geq n > m} Cx_n y_m D.$$

In terms of the latter notation, the conclusion of our lemma under item (a) for the finite sequences X_p and Y_p, is equivalent to

(1) $\quad C^{-1}L_p D^{-1} \cap C^{-1}R_p D^{-1} = \phi.$

We now establish the inductive step, that is choose x_{p+1} and y_{p+1} such that the lemma is valid for n,m,i,j in {1,2,...,p+1}.

Since S is not relatively neo-compact while both

$$T := C^{-1}(C(C^{-1}L_pD^{-1})D)(Y_pD)^{-1} \quad \text{and}$$

$$T' := C^{-1}(C(C^{-1}(CX_pY_pD)D^{-1})D)(Y_pD)^{-1}$$

are relatively neo-compact, we can choose x_{p+1} in $S\setminus(T\cup T')$.

Now, that $x_{p+1} \notin T$ is equivalent to

$$(2) \quad C^{-1}(Cx_{p+1}Y_pD)D^{-1} \cap C^{-1}L_pD^{-1} = \phi$$

while that $x_{p+1} \notin T'$ is equivalent to

$$(3) \quad C^{-1}(Cx_{p+1}Y_pD)D^{-1} \cap C^{-1}(CX_pY_pD)D^{-1} = \phi.$$

Also the subsets

$$\mathcal{Q} := (CX_{p+1})^{-1}\ (C(C^{-1}(CX_pY_pD)D^{-1})D)D^{-1} \quad \text{and}$$

$$\mathcal{Q}' := (CX_{p+1})^{-1}\ (C(C^{-1}(R_p \cup Cx_{p+1}Y_pD)D^{-1})D)D^{-1}$$

are relatively neo-compact, and so we can choose y_{p+1} in S such that $y_{p+1} \notin \mathcal{Q}$ and $y_{p+1} \notin \mathcal{Q}'$. Equivalently, this is such that

$$(4) \quad C^{-1}(CX_{p+1}y_{p+1}D)D^{-1} \cap C^{-1}(CX_pY_pD)D^{-1} = \phi \quad \text{and}$$

$$(5) \quad C^{-1}(CX_{p+1}y_{p+1}D)D^{-1} \cap C^{-1}(R_p \cup Cx_{p+1}Y_pD)D^{-1} = \phi$$

(respectively).

Now for the finite sequences X_{p+1} and Y_{p+1}, item (3) and the inductive hypothesis show that the lemma holds under condition (b), item (4) and the inductive hypothesis show that the lemma holds under condition (c), and to verify the lemma under condition (a) it is sufficient to establish item (1) with p+1 in place of p. To the latter end, we note that the inductive hypothesis, items (2) and (5) imply that

$$C^{-1}L_{p+1}D^{-1} \cap C^{-1}R_{p+1}D^{-1} = C^{-1}(L_p \cup CX_{p+1}y_{p+1}D)D^{-1} \cap C^{-1}(R_p \cup Cx_{p+1}Y_pD)D^{-1}$$

$$= (C^{-1}L_pD^{-1} \cap C^{-1}(R_p \cup Cx_{p+1}Y_pD)D^{-1})$$

$$(C^{-1}(CX_{p+1}y_{p+1}D)D^{-1} \cap C^{-1}(R_p \cup Cx_{p+1}Y_pD)D^{-1})$$

$$= \phi.$$

Repeating the argument countably many times we get our lemma. This completes the proof.

Remark. We warn the reader that in general conditions (a), (b) and (c) together, are weaker and not equivalent to the condition (d): $(n,m) \neq (i,j)$. It remains an open problem whether one can establish the conclusion of our lemma under (the stronger) condition (d). In fact even for a topological group S, this is not known to us. We strongly believe that if one can supply an affirmative answer to this problem, for a locally compact topological group S, then property (E) mentioned by Ching Chou [18] may be dropped in most of the results of [18].

Next we introduce the role of absolutely continuous measures in the following lemma.

3.2 LEMMA. *Let* S *be a* C-*distinguished topological semigroup,* ν *and* μ *measures in* M(S) *such that the maps* $x \to \nu * \bar{x}$ *and* $x \to \bar{x} * \mu$ *of* S *into* M(S) *are weakly continuous. Then*

$\nu o h o \mu \in WUC(S)$ *for all* $h \in M(S)^*$.

Proof. We first note that for all $\lambda \in M(S)$ and $h \in M(S)^*$, $\nu o h o \lambda \in C(S)$, since $\nu o h o \lambda(x) = h \circledast \lambda(\nu * \bar{x})$ and the map $x \to \nu * \bar{x}$ is weakly continuous. To show that $\nu o h o \mu$ is in RWUC(S), for example, we take a $\zeta \in C(S)^*$ and must show that the function $x \to \zeta((\nu o h o \mu)_x)$ is continuous. But this follows from the fact that $(\nu o h o \mu)_x = \nu o h o \bar{x} * \mu$ and the functional $\lambda \to \zeta(\nu o h o \lambda)$ is in $M(S)^*$. Similarly $\nu o h o \mu \in LWUC(S)$ and we are done.

We now prove the main result of this section.

3.3 THEOREM. *Let S be a C-distinguished topological semigroup admitting a non-zero absolutely continuous measure* (i.e. $M_a(S)$ is non-zero). *Then if S is not relatively neo-compact, we have that the quotient space* $WUC(S)/_{WAP(S)}$ *contains a linear isometric copy of* ℓ^∞ *and so is nonseparable.*

Proof. As noted in item 1.5, we trivially have $WAP(S) \subseteq WUC(S)$.

Since $M_a(S)$ is solid (by Theorem 2.1.7), we can choose a positive measure $\eta \in M_a(S)$ such that $\|\eta\| = 1$ and $K := \mathrm{supp}(\eta)$ is compact. By Corollary 2.1.13 we can find $u,v \in S$ such that if $\nu := \eta*\bar{u}$ and $\mu := \bar{v}*\eta$ then the maps $x \to \nu*\bar{x}$ and $x \to \bar{x}*\mu$ of S into M(S) are weakly continuous. We keep these measures, ν and μ, fixed for the remainder of our proof and note that both $C_o := \mathrm{supp}(\nu)$ and $D_o := \mathrm{supp}(\mu)$ are compact, (in fact $C_o = Ku$ and $D_o = vK$), and $\|\nu\| = \|\mu\| = 1$. Let $C := C_o \cup \{1\}$ and $D := D_o \cup \{1\}$, where 1 is the identity of S (if there is one) or an adjoined isolated identity of S.

Let sequences $A := \{x_1, x_2, \ldots\}$ and $B := \{y_1, y_2, \ldots\}$ be chosen as in Lemma 3.1 with respect to the compact sets C and D. We can choose infinite subsequences $A_k := \{x_{k_1}, x_{k_2}, \ldots\}$ of A and $B_k := \{y_{k_1}, y_{k_2}, \ldots\}$ of B such that

(α) $\bigcup_{k=1}^{\infty} A_k \subset A$ and $\bigcup_{k=1}^{\infty} B_k \subset B$;

(β) if $n \neq m$ then $A_n \cap A_m =$ and $B_n \cap B_m = \phi$;

(γ) Lemma 3.1 remains valid with k_n, k_m, k_i and k_j in place of n,m,i and j, respectively (i.e. when A_k and B_k replace A and B, respectively).

Let

$$E_k := \bigcup_{i=1}^{\infty} \bigcup_{i \leq j} Cx_{k_i}y_{k_j}D \quad \text{and} \quad F_k := \bigcup_{j=1}^{\infty} \bigcup_{i > j} Cx_{k_i}y_{k_j}D.$$

We define the functions f_k on S by

$$f_k(x) := \nu*\bar{x}*\mu(E_k) - \nu*\bar{x}*\mu(F_k).$$

Let $\{c_k\}$ be any element in ℓ^∞.

We now show that $\sum_{k=1}^{\infty} c_k f_k$ is in WUC(S). From (β) and (γ) we note that all the E_k's and F_k's are pairwise disjoint and so can define the functional $h \in M(S)^*$ by

$$h(\eta) := \sum_{k=1}^{\infty} c_k(\eta(E_k) - \eta(F_k)).$$

Now a simple exercise on our definitions shows that

$$\nu o h o \mu = \sum_{k=1}^{\infty} c_k f_k$$

and so $\sum_{k=1}^{\infty} c_k f_k \in WUC(S)$ by Lemma 3.2.

It remains to show the (clearly linear) map

$$\{c_k\} \to \sum_{k=1}^{\infty} c_k f_k + WAP(S)$$

is an isometry of ℓ^{∞} into $WUC(S)/_{WAP(S)}$. To achieve this we start by demonstrating that

(1) $$\|f+g\|_S \geq \|\{c_k\}\|_{\infty} ,$$

where $f := \sum_{k=1}^{\infty} c_k f_k$ and $g \in WAP(S)$. Suppose on the contrary there exists a g in WAP(S) and $\varepsilon > 0$ such that

$$\|f+g\|_S < \|\{c_k\}\|_{\infty} - 2\varepsilon.$$

We can find a positive integer k' such that $\|\{c_k\}\|_{\infty} - \varepsilon < |c_{k'}|$. We may assume that $c_{k'}$ is non-negative and hence obtain

(2) $$\|f+g\|_S < c_{k'} - \varepsilon.$$

From the definition of $f := \sum_{k=1}^{\infty} c_k f_k$ we have that

$$(3) \quad f(x_{k'_i}, y_{k'_j}) = c_{k'} f_{k'}(x_{k'_i}, y_{k'_j}) = \begin{cases} c_{k'} & \text{if } i \leq j \\ -c_{k'} & \text{if } i > j . \end{cases}$$

From (2) and (3) we obtain

$$(4) \quad \begin{cases} i \leq j & \text{implies } g(x_{k'_i}, y_{k'_j}) \leq -c_{k'} + |c_{k'} + g(x_{k'_i}, y_{k'_j})| < -\varepsilon \\ i > j & \text{implies } g(x_{k'_i}, y_{k'_j}) \geq c_{k'} - |-c_{k'} + g(x_{k'_i}, y_{k'_j})| > \varepsilon. \end{cases}$$

From (4) and Proposition 1.2, we have that $g \notin WAP(S)$, which contradicts our original choice of g in WAP(S). Hence (1) holds.

Noting that

$$\| \sum_{k=1}^{\infty} c_k f_k \|_S = \| \{c_k\} \|_\infty$$

and recalling (1), we have that

$$\| \sum_{k=1}^{\infty} c_k f_k + WAP(S) \|_{WUC(S)/WAP(S)} := \inf\{ \| \sum_{k=1}^{\infty} c_k f_k + g \|_S : g \in WAP(S)\}$$

$$= \| \{c_k\} \|_\infty .$$

Consequently the mapping $\{c_k\} \to \sum_{k=1}^{\infty} c_k f_k + WAP(S)$ is a linear isometry of ℓ^∞ into $WUC(S)/_{WAP(S)}$ and our proof is complete.

3.4 COROLLARY. *Let* S *be a non relatively neo-compact stip with* $M_a(S)$ *non-zero. (In particular* S *may be any locally compact group which is not compact.) Then the quotient space* $UC(S)/_{WAP(S)}$ *contains an isometric linear copy of* ℓ^∞.

Proof. We have UC(S) = WUC(S), by Theorem 1.5(i), and so our result follows from Theorem 3.3.

Our next result and Theorem 3.7 are versions of Theorem 3.3 with proofs essentially contained in that of Theorem 3.3.

3.5 PROPOSITION. *Let* S *be a* C*-distinguished topological semigroup,* ρ *a positive measure in* $M_a(S)$ *with compact support,* $\|\rho\| = 1$ *and suppose that* S *is not relatively neo-compact. Then* $WUC(S)o\rho \setminus WAP(S)$ *is non-separable (in* C(S)*).*

Proof. Let $F := \text{norm-cl}(WUC(S)o\rho)$, $A := F \cap WAP(S)$ and $E := \text{supp}(\rho)$. In the proof of Theorem 3.3 take $D := (E\cup\{1\})(D_o\cup\{1\})$ and let the functions f_k be as constructed there. Then the (clearly linear) map

$$\{c_k\} \to \sum_{k=1}^{\infty} c_k(f_k o\rho) + A$$

is an isometry of ℓ^∞ into F/A and hence $F \setminus A$ is non-separable. This implies our result.

For the purpose of our next lemma, we recall that if A is any set of functionals then $A^+ := \{a \in A : a \geq 0\}$ and $A^- := \{a \in A : a \leq 0\}$.

3.6 LEMMA. *For any* $\eta \in M(S)$, *let* $\eta^\perp := \{h \in M(S)^* : h(\nu*\eta) = 0$ *for all* $\nu \in M_a(S)\}$. *Then if* ρ *and* η *are any positive measures in* M(S) *with* $\rho \ll \eta$, *we have that*

$$(\eta^\perp)^+ \subseteq (\rho^\perp)^+.$$

Proof. Let $h \in (\eta^\perp)^+$ and $\nu \in M_a(S)$. We may assume that $\nu = \nu_1 - \nu_2$, where ν_1 and ν_2 are positive measures in $M_a(S)$. Then for i = 1 or 2,

$$h(\nu_i*\eta) = \int h(\nu_i*\bar{x})\,d\eta(x) = 0$$

and so $h(\nu_i*\bar{x}) = 0 \quad (x \in \text{supp}(\eta))$ since $x \to h(\nu_i*\bar{x})$ is a non-negative continuous function. Since $\rho \ll \eta$ we have $\text{supp}(\rho) \subseteq \text{supp}(\eta)$ and so

$h(\nu_i*\rho) = \int h(\nu_i*\bar{x})\,d\rho(x) = 0.$ So we conclude that

$$h(\nu*\rho) = h(\nu_1*\rho) - h(\nu_2*\rho) = 0$$

and $h \in (\rho^\perp)^+$.

3.7 THEOREM. *Let S be a C-distinguished topological semigroup, η a positive measure in $M_a(S)$ with $\|\eta\| = 1$, and suppose that*

$$WUC(S) \subseteq \text{norm-cl}(WAP(S) + \eta^{\perp} + X)$$

for some norm separable $X \subseteq M_a(S)^*$. Then S is relatively neo-compact.

Proof. Let $\mu \in M_a(S)$ be a positive measure with compact support and such that $\mu << \eta$ and $\|\mu\| = 1$. By Lemma 3.6,

$$WUC(S)^+ \subseteq \text{norm-cl}(WAP(S)^+ + (\mu^{\perp})^+ + X_1)$$

for some norm separable $X_1 \subseteq M_a(S)^*$. For each $h \in (\mu^{\perp})^+$ and $\nu \in M(S)$, we have $\nu*\mu \in M_a(S)$, so $h \odot \mu^2(\nu) := h(\nu*\mu^2) = h((\nu*\mu)*\mu) = 0$. Consequently $(\mu^{\perp})^+ \odot \mu^2$ is zero and, taking $\rho := \mu^2$, we get

$$WUC(S)^+ \odot \rho \subseteq \text{norm-cl}(WAP(S)^+ \odot \rho + X_1 \odot \rho) \subseteq \text{norm-cl}(WAP(S) + X')$$

for some norm separable $X' \subseteq M_a(S)^*$. Similarly, we can find a norm separable $X'' \subseteq M_a(S)^*$ such that

$$WUC(S)^- \odot \rho \subseteq \text{norm-cl}(WAP(S) + X'').$$

Hence $WUC(S) \odot \rho \setminus WAP(S)$ is separable (in the norm of $M_a(S)^*$).

Now the norm of $M_a(S)^*$ is in fact $\| \ \|_{P(M_a(S))}$; so for all f and g in WUC(S), noting that $\bar{x}*\rho \in P(M_a(S))$ whenever $x \in S$, we thus have

$$\|f\circ\rho^2 - g\circ\rho^2\|_S := \sup\{|f(\bar{x}*\rho^2) - g(\bar{x}*\rho^2)| : x \in S\}$$

$$\leq \sup\{|f\odot\rho(\nu) - g\odot\rho(\nu)| : \nu \in P(M_a(S))\} = \|f\odot\rho - g\odot\rho\|_{P(M_a(S))}.$$

Therefore $WUC(S)\circ\rho^2 \setminus WAP(S)$ is separable with respect to the norm of C(S). By Proposition 3.5, it follows that S is relatively neo-compact.

In the case of groups from Theorems 3.7 and 1.5, we have the following improvement of a result of E.E. Granirer [50].

3.8 COROLLARY. *Let* G *be any locally compact topological group and* $\alpha \in L^1(G)$, $\alpha \geq 0$ *and* $\|\alpha\| = 1$. *Let* $\alpha^{\perp} := \{f \in L^{\infty}(G) : (f, L^1(G)*\alpha) = 0\}$. *If* $UC(G) \subseteq$ *norm-cl*$(WAP(G) + \alpha^{\perp} + X)$ *for some norm separable* $X \subseteq L^{\infty}(G)$, *then* G *is compact.*

For ease of reference we mention the following consequence of Thoerem 3.3.

3.9 COROLLARY. *Let* S *be a* C-*distinguished topological semigroup with* $M_a(S)$ *non-zero and such that* $C^{-1}D$ *and* DC^{-1} *are compact for all compact subsets* C *and* D *of* S. *Then* S *is locally compact and the following items are equivalent:*

(i) S *is compact;*
(ii) $C(S) = WAP(S)$;
(iii) $WUC(S) = WAP(S)$.

3.10 Notes. Let G be a locally compact topological group. Burckel [13] proved that C(G) = WAP(G) if and only if G is compact. Then E.E. Granirer [49] improved this to: UC(G) = WAP(G) if and only if G is compact. Generalizing these results Dzinotyiweyi [35] proved Corollary 3.9. For an amenable group (i.e. a group G with UC(G) admitting an invariant mean), Corollaries 3.4 and 3.8 were proved by Granirer in [49] and [50], respectively. Also Ching Chou [18] proved Corollary 3.4 for groups satisfying a certain condition he calls property (E). The generalization of these results given in Theorems 3.3 and 3.7 is taken from Dzinotyiweyi [37].

It is natural to ask what happens for the situation where $M_a(S)$ is zero. For instance, a close examination of Corollary 3.9 leads us to the following conjecture raised in [36].

CONJECTURE A. Let S be a locally compact semi-topological semigroup such that $C^{-1}D$ and DC^{-1} are compact for all compact subsets C and D of S. Then the following items are equivalent:

(a) S is pseudocompact,
(b) C(S) = WAP(S),
(c) WUC(S) = WAP(S).

Evidently (b) implies (c) and below we show that (a) implies (b). Thus the conjecture will be solved if one can show that (c) implies (a). However, (if this turns out to be difficult), it would also be interesting if one could show that (b) implies (a).

PROPOSITION. *We have that* (a) *implies* (b).

Proof. It is a simple exercise to show that every locally compact pseudocompact space is countably compact. (See e.g. Gillman and Jerison [44].) This together with Lemma 4.2 implies our result.

We also have

THEOREM [36]. *If in addition to the hypothesis of our conjecture* S *is* σ*-compact, then* C(S) = WAP(S) *if and only if* S *is compact.*
(In fact the actual result proved in [36] is more general.)

Thus our conjecture is known for the case where S is a topological semigroup admitting a non-zero absolutely continuous measure (see Corollary 3.9) or S is σ-compact. In general item (b) does not imply S is compact as we note in the following example.

EXAMPLE [35]. Recalling the usual well-ordering of the set of all ordinals and taking ω_1 to be the first uncountable ordinal, let

$$S := \{\alpha : \alpha \text{ is an ordinal and } \alpha < \omega_1\}$$

with the usual interval topology and maximum operation. Now one can note that

(i) S is a locally compact topological semigroup with $C^{-1}D$ and DC^{-1} compact for all compact subsets C and D of S;

(ii) S is pseudocompact but not compact (see e.g. [44]);

(iii) $M_a(S)$ is zero;

(iv) C(S) = WAP(S), (by (ii) and the above Proposition).

We have already noted that a stip S behaved as if $M_a(S)$ were non-zero. So the following conjecture seems reasonable.

CONJECTURE B [37]. If S is a stip that is not relatively neo-compact, the quotient space $UC(S)/_{WAP(S)}$ contains an isometric linear copy of ℓ^∞.

Even if it turns out that every stip S is such that $M_a(S)$ is non-zero, it would be interesting to solve this conjecture without the use of $M_a(S)$. We suspect that the notion of functions with separable orbits may be useful in finding a function in $UC(S)\setminus WAP(S)$.

4. THE STONE-ČECH COMPACTIFICATION OF A SEMIGROUP

In this section we take S *to be any completely regular semitopological semigroup.*

We now define the Stone-Čech compactification βS of S. For each $f \in C(S)$, let $D_f := \{\lambda \in \mathbb{C} : |\lambda| \leq \|f\|_S\}$ and note that D_f is a compact disc. Hence, by Tychonoff's theorem, $E := \times\{D_f : f \in C(S)\}$ is compact. In a natural manner we embed S into E via the mapping $\theta : S \to E$ given by

$$\theta(x)(f) := f(x) \quad \text{for all} \quad x \in S \quad \text{and} \quad f \in C(S).$$

Then $\beta S := cl(\theta(S))$. For our convenience we shall write x in place of $\theta(x)$ (and hence S in place of $\theta(S)$), for all x in S.

Suppose that S is locally compact. Then $C(S)^*$ is the space of bounded finitely additive regular Borel measures on S. It is interesting to study $C(S)^*$ as a Banach algebra with an Arens or convolution product. This idea is suggested in Hewitt and Ross [59, page 275] and has been studied to some extend by Butcher [14], Olubummo ([75] and [76]) and Pym and Vasudema ([84] and [85]). Now, since C(S) can be identified with $C(\beta S)$, the Riesz representation theorem enables us to identify $C(S)^*$ with $M(\beta S)$. So one may obtain some information about $C(S)^*$ by studying βS as a semigroup.

First we introduce an "Arens operation" on βS: For each $\alpha \in \beta S$ and $f \in C(S)$, we define the function $\alpha o f$ on S by $\alpha o f(x) := \alpha({}_x f)$ for all x in S. For all $\alpha, \beta \in \beta S$ we define the Arens operation, o, by

$$\alpha o \beta(f) := \alpha(\beta o f) \quad \text{for all} \quad f \in C(S).$$

It is clear that this operation is well defined if and only if $\beta o f \in C(S)$ for all $\beta \in C(S)$ and $f \in C(S)$.

Let $LMC(S) := \{f \in C(S) : \alpha o f \in C(S) \text{ for all } \alpha \in \beta S\}$. Then we have the following simple exercise for the reader.

4.1 PROPOSITION. *The following items are equivalent:*

(i) $(\beta S, o)$ *is a left semitopological semigroup;*
(ii) $LMC(S) = C(S)$.

Next we are interested in the topological structure of semigroups S for which $(\beta S, o)$ is a left semitopological semigroup. Partly towards this end we include the following lemma.

4.2 LEMMA. *If* S *is countably compact then* $C(S) = WAP(S)$.

Proof. Let $\{x_n\}$ and $\{y_m\}$ be sequences in S. If S is countably compact, we can find cluster points x and y in S of $\{x_n\}$ and $\{y_m\}$, respectively. So for each $f \in C(S)$ we clearly have $f(xy)$ in the closure of each of the sets $\{f(x_n y_m) : n < m\}$ and $\{f(x_n y_m) : n > m\}$. Recalling Proposition 1.2 we thus have $f \in WAP(S)$ and our result follows.

4.3 THEOREM. *Let* S *be (topologically) normal, locally compact, right cancellative and such that* $C^{-1}D$ *is relatively compact for all compact subsets* C *and* D *of* S. *Then the following items are equivalent:*

(i) $(\beta S, o)$ *is a left semitopological semigroup;*
(ii) S *is discrete or countably compact.*

Proof. From our definitions and lemma 4.2, it is trivial to note that item (ii) implies that $LMC(S) = C(S)$ and hence item (i) follows by Proposition 4.1.

Now suppose that (i) holds with S neither discrete nor countably compact. We can find a relatively compact infinite set $\{s_n : n \in \mathbb{N}\}$ in S and a sequence $\{U_n\}$ of subsets of S such that U_n is a neighbourhood of s_n and $s_m \notin U_n$ for $m \neq n$. Let $C := cl(\{s_n : n \in \mathbb{N}\})$ and note that C is compact. We can choose a sequence $\{t_n\}$ in S without a cluster point and such that

$$t_{n+1} \notin \bigcup_{i=1}^{n} C^{-1}(Ct_i) \quad \text{for all} \quad n \in \mathbb{N}. \tag{1}$$

Next we consider the closure $\bar{X}$ of the set $X := \{s_m t_n : m,n \in \mathbb{N}\}$. Let $(s_{m(\alpha)} t_{n(\alpha)})$ be a set in X converging to $x \in S$. Let D be a compact neighbourhood of x and choose α_o such that $\alpha \geq \alpha_o$ implies that $s_{m(\alpha)} t_{n(\alpha)} \in D$.

Thus $t_{n(\alpha)} \in C^{-1}D$ for $\alpha \geq \alpha_o$. Since $C^{-1}D$ is relatively compact and $\{t_n\}$ has no cluster point, we must have $\{t_{n(\alpha)} : \alpha \geq \alpha_o\}$ finite. Hence $x = ct_n$ for some $n \in \mathbb{N}$ and $c \in C$. The representation $x = ct_n$ is unique, for if $ct_n = c_1 t_m$ for some $c_1 \in C$ and $m \in \mathbb{N}$, then $n = m$ by (1) and so $c = c_1$ by right cancellation.

It is now clear that if we define f on $\bar{X}$ by setting

$$f(s_m t_n) := \begin{cases} 1 & \text{if} \quad m < n \\ 0 & \text{if} \quad m \geq n \end{cases}$$

$$f(ct_n) := 0 \quad \text{otherwise} \quad (\text{for } c \in C),$$

then f is continuous. By the Tietze extension theorem, f extends to a continuous function, g say, in C(S). Hence, by Theorem 4.4, we have that $g \notin LMC(S)$ and so (i) does not hold, by Proposition 4.1. By this contradiction our proof is complete.

Since LMC(S) plays a significant role in the study of βS as a semigroup, the following characterizations of LMC(S) are of interest in this section.

First we recall that a (Hausdorff) topological space X is a k_o-space if every complex-valued function on X that is continuous on compact subsets of X is continuous on X. (Note that a k_o-space is not necessarily a k-space - see e.g. Exercise 4.13.)

If $\phi \in C(S)^*$ is such that $\phi(f.g) = \phi(f)\phi(g)$ we say ϕ is *multiplicative*.

4.4 THEOREM. *Let* S *be a* k_o*-space and* $f \in C(S)$. *Then the following items or equivalent:*

(i) $f \in LMC(S)$.

(ii) *For all sequences* $\{x_n\}$ *and* $\{y_n\}$ *in* S *with* $\{x_n : n \in \mathbb{N}\}$ *relatively compact, we have the closures of the sets* $\{f(x_n y_m) : n < m\}$ *and* $\{f(x_n y_m) : n > m\}$ *not disjoint.*

(iii) $\{{}_x f : x \in K\}$ *is weakly compact for every compact* $K \subseteq S$.

(iv) $f \in LWUC(S)$.

Proof. To show that (i) implies (ii), let $K := cl(\{x_n : n \in \mathbb{N}\})$ and let P be the set of all multiplicative means on $C(S)$ endowed with the weak*-topology. Thus both K and P are compact. For $f \in LMC(S)$, we have the map $F : K \times P \to \mathbb{C}$, given by

$$F(x,\eta) := \eta({}_x f) \quad \text{for all} \quad (x,\eta) \in K \times P,$$

separately continuous.

Assuming the notation of (ii), let x' be a cluster point of $\{x_n\}$ in K and ν a cluster point of $\{\bar{y}_m\}$ in P. Consequently every neighbourhood of $\nu({}_{x'}f)$ meets infinitely many columns and infinitely many rows of the double sequence $\{F(x_n,\bar{y}_m)\}$. So the closures of the sets $\{{}_{x_n}f(y_m) : n < m\}$ and $\{{}_{x_n}f(y_m) : n > m\}$ in $\mathbb{C}$ are not disjoint. Thus (i) implies (ii).

The equivalence of (ii) and (iii) follows from Grothendieck's Theorem (- see Appendix B.7).

That (iii) and (iv) are equivalent is left as a simple exercise for the reader.

Now from Theorem 4.4 and the proof of Theorem 4.3 we note that if a semigroup S satisfies the hypothesis of Theorem 4.3, then $C(S) = LWUC(S)$ if and only if S is either discrete or countably compact. We now show that an easy adjustment of the proof of Theorem 4.3 gives a more general result of independent interest.

4.5 THEOREM. *Let S be normal, locally compact and right cancellative. Suppose S is neither countably compact nor discrete and $C^{-1}D$ is compact for all compact subsets C and D of S. Then, for some closed subset $\bar{X}$ of S, we have that $(C(S)\setminus LWUC(S))_{|\bar{X}}$ contains a linear isometric copy of ℓ^∞ and so the quotient space $C(S)/_{LWUC(S)}$ is non-separable.*

Proof. We urge the reader to first study the proof of Theorem 4.3 as we will omit closely related details in this proof.

Let the sequences $\{s_n\}$ and $\{t_n\}$ and the set C be as constructed in the proof of Theorem 4.3. Choose infinite subsequences $T_k := \{t_{k_1}, t_{k_2}, \ldots\}$ of $T := \{t_1, t_2, \ldots.\}$ such that

$$\bigcup_{k=1}^{\infty} T_k = T, \text{ and}$$

$$T_k \cap T_{k'} = \phi \text{ if and only if } k \neq k'.$$

Let $X_k := \{s_m t_{k_n} : m,n \in \mathbb{N}\}$, $X := \{s_m t_n : m,n \in \mathbb{N}\}$ and note that our construction of the T_k's and T imply

(a) $\bar{X}_k = \{ct_{k_n} : c \in C \text{ and } n \in \mathbb{N}\}$,

(b) $\bar{X}_k \cap \bar{X}_{k'} = \phi$ if and only if $k \neq k'$,

(c) $\bigcup_{k=1}^{\infty} \bar{X}_k = \bar{X}$. (To verify these items, see proof of Theorem 4.3 for relevant techniques.)

Next we define the functions $f_k : \bar{X}_k \in \mathbb{R}$ by

$$f_k(s_m t_{k_m}) := \begin{cases} 1 & \text{if } m < n \\ -1 & \text{if } m \geq n \end{cases}$$

$$f_k(ct_{k_n}) := -1 \quad \text{if} \quad c \in C \setminus \{s_m : m \in \mathbb{N}\}.$$

Then (as similarly shown in the proof of Theorem 4.3) f_k is continuous, for all $k \in \mathbb{N}$. Corresponding to each element $\{d_{k'}\}$ in ℓ^∞, let $F_{(d_{k'})}$ be the function defined on $\bar{X}$ by

$$F_{(d_{k'})}(x) := d_\ell f_\ell(x) \quad \text{if and only if} \quad x \in \bar{X}_\ell \quad \text{for some } \ell \in \mathbb{N}.$$

By items (b) and (c) we thus have $F_{(d_{k'})}$ well defined as a function.

To see that $F_{(d_{k'})}$ is continuous, suppose $(c_\alpha t_{n(\alpha)})$ is a net in $\bar{X}$ converging to some point ct_{k_ℓ}, for some c_α, $c \in C$ and $t_{n(\alpha)}$, $t_{k_\ell} \in T$. Then, by the definition of T, we have $n(\alpha) = k$ eventually and so, eventually $(c_\alpha t_{n_\alpha}) \subseteq \bar{X}$. Consequently, eventually

$$F_{(d_{k'})}(c_\alpha t_{n(\alpha)}) = d_\ell f_\ell(c_\alpha t_{n(\alpha)}) \to d_\ell f_\ell(ct_{k_\ell}) = F_{(d_{k'})}(ct_{k_\ell}).$$

Thus $F_{(d_{k'})}$ is continuous.

Now noting that

$$(*) \quad F_{(d_{k'})}(s_m t_{k_n}) = \begin{cases} d_k & \text{if} \quad m < n \\ -d_k & \text{if} \quad m \geq n \end{cases},$$

Theorem 4.4 and Tietze's Extension Theorem imply the existence of a function $\bar{F}_{(d_{k'})} \in C(S)\backslash LWUC(S)$ such that

$$\bar{F}_{(d_{k'})}|\bar{X} = F_{(d_{k'})} \quad \text{and} \quad \|\bar{F}_{(d_{k'})}\|_S = \|F_{(d_{k'})}\|_{\bar{X}} = \|\{d_{k'}\}\|_\infty .$$

Thus the (clearly) linear map $\{d_{k'}\} \to \bar{F}_{(d_{k'})}|\bar{X}$ of ℓ^∞ into $(C(S)\backslash LWUC(S))_{|\bar{X}}$ is isometric.

Since ℓ^∞ is non-separable it follows that $C(S)\backslash LWUC(S)$ and hence $C(S)/_{LWUC(S)}$ is non-separable.

Theorems 4.3 and 4.4 are taken from Baker and Butcher [7]. The argument used in showing that (i) implies (ii) in the proof of Theorem 4.4 is inspired by the proof of [106, Theorem 1].

Also in the paper [7], Baker and Butcher showed that βS is a left semi-topological semigroup under a certain operation extending that of S while $C(S) \neq LMC(S)$, for many locally compact semitopological semigroups S. Such arbitrary products unfortunately do not seem to reflect much on the structure of βS or LMC(S). On the other hand the Arens type of products seem to be of interest. In what follows, we make use of Arens products defined with respect to various subspaces of C(S).

4.6 NOTATION. For the remainder of this section let $A_S \in \{LMC(S), LWUC(S), WUC(S), C(S)\}$ and A(S) be the compact space constructed with respect to A_S as we constructed βS with respect to C(S). (Thus A(S) is the spectrum of the algebra A_S.)

4.7 EXERCISES. For all $f \in A_S$ and $\alpha \in A(S)$, let αof be the function on S given by $\alpha of(x) := \alpha({}_x f)$ for all $x \in S$. Then if $A_S \subseteq LMC(S)$ we have

(i) $\alpha of \in A_S$

(ii) $(A(S),o)$ is a left semitopological semigroup with the Arens operation 'o' given by $\alpha o\beta(f) := \alpha(\beta of)$, for all $\alpha,\beta \in A(S)$.

4.8 PROPOSITION. *The following items are equivalent:*

(i) $A(S)$ *is a semitopological semigroup with subsemigroup* S;

(ii) $\mathcal{A}_S = WAP(S)$.

Proof. That (i) implies (ii) follows from our definition of A(S) and Lemma 4.2. Now suppose (ii) holds and consider the semigroup $(A(S),o)$ defined in Exercise 4.7(ii). To verify item (i) it is sufficient to show that $(A(S),o)$ is a right semitopological semigroup. To this end, let $f \in \mathcal{A}_S$ be fixed and suppose (α_γ) is a net converging to α in A(S). We have C_f := weak-closure of $\{_xf : x \in S\}$ compact and $(\alpha_\gamma of) \subseteq C_f$. Hence $(\alpha_\gamma of)$ has a weak cluster point g (say) in $\mathcal{A}_S$. In particular for all x in S we have that

$$\alpha_\gamma of(x) := \alpha_\gamma(_xf) \to \alpha(_xf) = \alpha of(x)$$

and g(x) is a cluster point of $(\alpha_\gamma of(x))$. Hence $g = \alpha of$. It follows that $\alpha_\delta of \to \alpha of$ in the $\sigma(\mathcal{A}_S,\mathcal{A}_S^*)$-topology. So for all $\beta \in A(S)$, we have

$$\beta o\alpha_\gamma(f) := \beta(\alpha_\gamma of) \to \beta(\alpha of) = \beta o\alpha(f).$$

Thus $(A(S),o)$ is a right semitopological semigroup.

4.9 THEOREM. *Let* S *be locally compact and such that* $C^{-1}D$ *and* DC^{-1} *are compact for all compact subsets* C *and* D *of* S. *If* S *is either* σ*-compact and* $\mathcal{A}_S = C(S)$ *or* S *is a topological semigroup supporting a non-zero absolutely continuous measure, we have the following items equivalent:*

(i) A(S) *is a semitopological semigroup with subsemigroup* S;

(ii) S *is compact.*

Proof. This is an immediate consequence of Corollary 3.9, the Theorem mentioned in item 3.10 and Proposition 4.8.

The advantage of the compactification A(S) of S is that already, if $\mathcal{A}_S \in \{LMC(S), LWUC(S), WUC(S)\}$, A(S) is a left semitopological semigroup. Our preceding Theorem now says that, for a large class of non-compact S,

A(S) is not a semitopological semigroup. A natural question to ask is: What is the largest semitopological semigroup $B_A(S)$ contained in (A(S),o) and such that S is a subsemigroup of $B_A(S)$? To formalize this question we first introduce the following notation. Let

$$B_A(S) := \{\beta \in A(S) : \text{the map } \alpha \to \beta o \alpha \text{ of } A(S) \text{ into } A(S) \text{ is continuous}\}.$$

$B_A(S)$ is called the set of *bicontinuous points* in A(S) in [88] (where S is a group).

4.10 EXERCISE. We always have $S \subseteq B_A(S)$.

The following conjecture seems reasonable.

4.11 CONJECTURE. Let S be locally compact and such that $C^{-1}D$ and DC^{-1} are compact for all compact subsets C and D of S. Then if S is either σ-compact or such that S is a topological semigroup admitting a non-zero absolutely continuous measure, we have $B_A(S) = S$.

For the case of a discrete topological abelian group, the above conjecture is due to Baker and Milnes [8]. We quote the following cases now known.

4.12 THEOREM (Ruppert [88]). *We have that* $B_A(S) = S$ *for each of the following cases:*

(i) S *is an abelian locally compact topological group;*
(ii) S *is a connected locally compact topological group;*
(iii) S *is a discrete topological group with the set of all elements in* S *of order two countable.*

We conclude this section by giving the reader an exercise promised before.

4.13 EXERCISE. In general every k-space is a k_o-space. We required S to be a k_o-space in Theorem 4.4. We now mention the following semitopological semigroup S which is a k_o-space but not a k-space.

Let τ be the topology on the real numbers with generic open sets of the type $O \setminus P$ with O open in the usual topology of $\mathbb{R}$ and P a countable subset of $\mathbb{R}$. Let $\mathbb{R}_\tau$ denote the resultant space. Then with the additive operation $S := \mathbb{R}_\tau$ is a semitopological semigroup that is a k_o-space but not a k-space. (See e.g. [9] for details.)

5. REGULARITY OF MULTIPLICATION IN SEMIGROUP ALGEBRAS

Let A be a Banach algebra with first and second dual spaces A^* and A^{**}, respectively. For $\phi \in A^{**}$, $h \in A^*$ and $\nu \in A$ we define $\nu o h$, $h o \nu$, h_ϕ and h^ϕ in A^* by

$$\nu o h(\mu) := h(\nu.\mu) \quad \text{and} \quad h o \nu(\mu) := h(\mu.\nu) \qquad (\mu \in A)$$

$$h_\phi(\mu) := \phi(h o \mu) \quad \text{and} \quad h^\phi(\mu) := \phi(\mu o h) \qquad (\mu \in A).$$

We then define the *Arens products* ([1] and [2]) $\circledcirc$ and $\circledcirc'$ on A^{**} by

$$\phi \circledcirc \psi(h) := \psi(h_\phi) \quad \text{and} \quad \phi \circledcirc' \psi(h) := \phi(h^\psi)$$

for all $\phi, \psi \in A^{**}$ and $h \in A^*$.

If the two Arens products coincide (i.e. $\phi \circledcirc \psi = \phi \circledcirc' \psi$ for all ϕ, ψ in A^{**}) we say multiplication in A is *regular*. Multiplication in A is said to be *irregular* if it is not regular. The products were first introduced by Arens and have been used in various publications. It is unusual to find a property of Banach algebras that is invariant under passage both to sub-algebras and to quotient algebras. One such invariant property is that of regularity of multiplication. In this section we are interested in finding when we have this property for algebras like $M_a(S)$.

With respect to either product $\circledcirc$ or $\circledcirc'$, A^{**} is a Banach algebra into which A is embedded isometrically by the canonical homomorphism $\nu \to \pi(\nu)$ $(\nu \in A)$ where

$$\pi(\nu)(h) := h(\nu) \qquad (h \in A^*).$$

It is immediate from the definitions that either Arens product on A^{**} is separately continuous for $\sigma(A^{**}, A^*)$. Any pair $\{\pi(\nu_n)\}$, $\{\pi(\mu_m)\}$ of bounded sequences in $\pi(A)$ will have cluster points ϕ, ψ (respectively) in A^{**} for $\sigma(A^{**}, A^*)$. If multiplication in A is regular, it follows that if $h \in A^*$ is such that both repeated limits of the double sequence $\{h(\nu_n.\mu_m)\}$ exist, then these two limits are equal (being in fact $\phi \circledcirc \psi(h)$).

Recalling Proposition 1.2, we summarise the above remarks (for measure algebras) in the following lemma.

5.1 LEMMA. *Let* S *be a* C-*distinguished topological semigroup and* A *a closed subalgebra of* M(S). *Then the following items are equivalent:*

(i) A *has regular multiplication;*

(ii) if $\{\nu_n\}$, $\{\mu_m\}$ *are any bounded sequences in* A *and* $h \in A^*$ *such that* $b := \lim_m \lim_n h(\nu_n * \mu_m)$ *and* $c := \lim_n \lim_m h(\nu_n * \mu_m)$ *exist, then* $b = c$;

(iii) *every* $h \in A^*$ *is a weakly almost periodic function on the topological semigroup* $(P(A), \|\ \|)$. (Recall the definition of P(A) from 2.1.)

For any $A \subseteq M(S)$ we recall that d_A denotes the density function as defined in item 1.3.2.

5.2 LEMMA. *Let* S *be a* C-*distinguished topological semigroup with identity element* 1 *and* A *a convolution measure algebra with foundation equal to* S. *Let* W *be a subset of* S *such that* 1 *is not isolated in* W *and* $W \subseteq d_A(S)$. *Then there exist sequences* $\{C_n\}$, $\{D_m\}$ *of non-A-negligible compact subsets of* W *such that, for all* $n,m,i,j \in \mathbb{N}$ *with* $n < m$ *and* $i > j$ *we have*

$$C_n D_n \cap C_i D_j = \phi.$$

Proof. Suppose, by the induction hypothesis, for some positive integer p we have finite sequences $\{C_1, C_2, \ldots, C_p\}$, $\{D_1, D_2, \ldots, D_p\}$ of non-A-negligible compact subsets of W such that if

$$X_p := \bigcup_{n=1}^{p} C_n, \quad Y_p := \bigcup_{n=1}^{p} D_n, \quad L_p := \bigcup_{n=1}^{p-1} \ \bigcup_{n<m\leq p} C_n D_m$$

and

$$R_p := \bigcup_{m=1}^{p-1} \ \bigcup_{p\geq n>m} C_n D_m$$

then

(α) $1 \notin X_p \cup Y_p \cup L_p \cup R_p$

(β) $X_p \cap Y_p = \phi$

(γ_1) $L_p \cap Y_p = \phi$

(γ_2) $R_p \cap X_p = \phi$

(δ) $L_p \cap R_p = \phi$

To prove the inductive step, we now choose C_{p+1} and D_{p+1} and verify items (α) to (δ) with p+1 in place of p. (Note that all sets mentioned in items (α) to (δ) are compact.)

We can choose a non-A-negligible compact set $C_{p+1} \subset W$ such that

(1) $1 \notin C_{p+1} \cup C_{p+1} Y_p$ by (α)

(2) $C_{p+1} \cap (Y_p \cup R_p) = \phi$ by (α)

(3) $C_{p+1}Y_p \cap (L_p \cup X_p \cup C_{p+1}) = \phi$ by (β), (γ_1) and since $1 \notin Y_p$.

Next we can choose a non-A-negligible compact set $D_{p+1} \subset W$ such that

(4) $1 \notin D_{p+1} \cup X_p D_{p+1}$ by (α)

(5) $D_{p+1} \cap (X_p D_{p+1} \cup L_p \cup X_{p+1}) = \phi$ by (α) and since $1 \notin C_{p+1}$.

(6) $X_p D_{p+1} \cap (C_{p+1} Y_p \cup R_p \cup Y_p \cup D_{p+1}) = \phi$ by (3), (γ_2), (β) and (α).

We are now in a position to establish items (α) to (δ) with p+1 in place of p.

Noting that $L_{p+1} = L_p \cup X_p D_{p+1}$ and $R_{p+1} = R_p \cup C_{p+1} Y_p$, it follows from ($\alpha$), (1) and (4) that

$$1 \notin X_{p+1} \cup Y_{p+1} \cup L_{p+1} \cup R_{p+1}.$$

From (β), (2) and (5) we get

$$X_{p+1} \cap Y_{p+1} = (X_{p+1} \cap Y_p) \cup (X_{p+1} \cap D_{p+1}) = \phi.$$

From (γ_1), (5) and (6) we get

$$L_{p+1} \cap Y_{p+1} = (L_p \cup X_p D_{p+1}) \cap Y_{p+1}$$
$$= (L_p \cap Y_{p+1}) \cup (X_p D_{p+1} \cap Y_{p+1}) = \phi.$$

From (δ), (3) and (6) we have

$$L_{p+1} \cap R_{p+1} = (L_p \cup X_p D_{p+1}) \cap (R_p \cup C_{p+1} Y_p)$$
$$= (L_p \cap R_p) \cup (L_p \cap C_{p+1} Y_p) \cup (X_p D_{p+1} \cap (R_p \cup C_{p+1} Y_p))$$
$$= \phi.$$

This completes our verification of the inductive step. By repeating the argument countably many times we note, in particular, that item (δ) leads to the conclusion of our lemma.

5.3 THEOREM. *Let* S *denote a* C*-distinguished topological semigroup with an identity element* 1 *and* A *a convolution measure algebra with foundation equal to* S. *Then*

(i) *If* 1 *is not isolated in* S *we have multiplication in* A *irregular.*

(ii) *If the foundation of* $M_a(S)$ *coincides with* S *and the sets* $x^{-1}\{y\}$ *and* $[y]x^{-1}$ *are finite* $(x,y \in S)$, *then* A *has regular multiplication if and only if* S *is finite.*

Proof. To prove (i) suppose that 1 is not isolated and choose sequences $\{C_n\}$, $\{D_n\}$ as in Lemma 5.2 (with $W := S$). Since A is solid, we can find positive measures ν_n, μ_n in A such that

$$\|\nu_n\| = \nu_n(C_n) = 1 \quad \text{and} \quad \|\mu_n\| = \mu_n(D_n) = 1 \qquad (n \in \mathbb{N}).$$

Since $L := \bigcup_{n=1}^{\infty} \bigcup_{n<m} C_n D_m$ is σ-compact, we can define $h \in A^*$ by

$$h(\eta) := \eta(H) \qquad (\eta \in A).$$

We then get

$$h(\nu_n * \mu_m) := \nu_n * \mu_m(L) = \begin{cases} 1 & \text{if} \quad n < m \\ 0 & \text{if} \quad n > m. \end{cases}$$

Recalling Lemma 5.1, our proof for item (i) is complete.

To prove item (ii) we first remark that if S is finite then trivially multiplication in A is regular. For the converse we first consider the case where S is discrete. Now if S is discrete and infinite, recalling Lemma 3.1(a) (with $C = D = \{1\}$), an argument similar to that used in the proof of item (i) (which we omit) easily shows that A has irregular multiplication. Thus S must be finite if A has regular multiplication. Finally we consider the case where S is not discrete. In view of item (i) our proof will be complete if we can show that 1 is not isolated. Since S is the foundation of $M_a(S)$, if 1 is isolated than $\bar{1} \in M_a(S)$ and S is discrete - see e.g. Exercise 2.3.10(c)(ii). By this conflict if S is not discrete, 1 is not isolated and our result follows.

5.4 THEOREM. *Let* A *be a convolution measure algebra whose foundation (semigroup)* A *is a topological subsemigroup of a topological group* G. *Then the following items are equivalent:*

(i) *Multiplication in* A *is regular*;

(ii) S *is finite.*

Proof. That (ii) implies (i) is trivial. Suppose (i) holds. We may assume that S is closed in G for otherwise we may take $\bar{S}$ in place of S and note that $\bar{S}$ is a closed subsemigroup which we may also take to be the foundation of A. If S is discrete, that (i) implies (ii) can be verified as done in Theorem 5.3(ii).

Suppose S is not discrete. If S is not compact then by arguing in a manner simpler to the proof of Lemma 3.1(a) we can find sequences $\{C_n\}$, $\{D_m\}$ of non-A-negligible compact subsets of S such that $n < m$ and $i > j$ imply that

$$C_nD_m \cap C_iD_j = \phi \quad (n,m,i,j \in \mathbb{N}).$$

Recalling our proof of Theorem 5.3(i), this similarly leads us to a contradiction. Next we suppose that S is compact. Then S is a compact group, by Theorem 1.2.3, and Theorem 5.3(ii) gives us the result.

5.5 <u>Notes</u>. (i) Civin and Yood [19] proved that if G is an infinite abelian locally compact topological group, multiplication in $L^1(G)$ is irregular. Then Young [105] improved the result to include the case where G is not necessarily abelian. Pym [81] studied the regularity of multiplication for certain convolution measure algebras supported on semigroups and so did Young [106]. Motivated by these results Dzinotyiweyi [35] proved the preceding two Theorems. In particular Theorem 5.4 is a generalization of Theorem 5 of Pym [81].

(ii) At this stage it should be clear to the reader that: the phenomenon of continuity of multiplication in βS, the weakly almost periodicity of functions on S and the regularity of multiplication of convolution measure algebras "living" on S, are somewhat related. Indeed if S_d denotes the semigroup S with discrete topology, then the reader may easily extract a proof for the following Theorem of Young [106] from Sections 3,4 and 5.

THEOREM. *The following are equivalent for any* C-*distinguished topological semigroup* S:

(a) M(S) *has regular multiplication;*

(b) $\ell^1(S)$ *has regular multiplication;*

(c) βS_d *admits the structure of a semitopological semigroup with* S *as a subsemigroup;*

(d) *there is no pair of sequences* $\{x_n\}$, $\{y_m\}$ *in* S *such that the sets* $\{x_n y_m : n < m\}$ *and* $\{x_n y_m : n > m\}$ *are disjoint;*

(e) $m(S) = WAP(S_d)$.

(iii) A major difference between Lemma 3.1(a) and Lemma 5.2 is that; in the former sequences $\{x_n\}$, $\{y_m\}$ are chosen in a "scattered manner" where as in Lemma 5.2 the sequences $\{C_n\}$, $\{D_m\}$ are chosen within a given "vicinity" of the identity element of S.

Now let S := [0,1] with maximum operation. Then $S = 1^{-1}\{1\}$ and S_d is relatively neo-compact. However, from Lemma 5.2 and the above Theorem, we have that βS_d is not a semitopological semigroup with subsemigroup S. Thus the Stone-Cech compactification of a locally compact relatively neo-compact topological semigroup is not necessarily a "nice" semitopological semigroup.

(iv) THE RADICAL OF $M_a(S)^{**}$. Throughout this item S denotes a foundation semigroup with identity element 1 (unless otherwise stated) and $\mathcal{R}_a(S)$ the radical of the Banach algebra $M_a(S)^{**}$ with Arens multiplicaton $\odot'$. For convenience we assume all spaces to be real. We are interested in showing that, for many cases, the space $\mathcal{R}_a(S)$ is very large. Towards this end we first prove the following result, also of independent interest.

PROPOSITION. *Let* S *be non-discrete and right cancellative. Then the quotient spaces* $M_a(S)^*/C(S)$ *and* $M_a(S)^*/LWUC(S)$ *contain isometric linear copies of* ℓ^∞.

Proof. Let W be a compact neighbourhood of 1 and corresponding to each function g in C(S) let G be the function in C(W×W) given by

$$G(x,y) := g(xy) \quad \text{for all} \quad x,y \in W.$$

Then a simple compactness argument shows that the set $\{G(x,.) : x \in W\}$ is relatively (norm and hence) weakly compact in $C(W)$.

We can find a sequence $\{V_k\}$ of disjoint open neighbourhoods contained in W. Choose $v_k \in V_k \cap S_1$ and recall that $V_k v_k^{-1}$ is a neighbourhood of 1. So there is a sequence $\{U_k\}$ of open neighbourhoods of 1 such that $U_k^2 \subseteq V_k v_k^{-1}$, for all $k \in \mathbb{N}$. By Lemma 5.2 we can choose sequences $\{C_{k_n}\}$, $\{D_{k_n}\}$ of non-$M_a(S)$-negligible compact subsets of U_k such that, for all $n,m,i,j \in \mathbb{N}$, we have

$$C_{k_n} D_{k_m} \cap C_{k_i} D_{k_j} = \phi \quad \text{whenever } n < m \text{ and } i > j.$$

By right cancellation we have

$$C_{k_n} D_{k_m} v_k \cap C_{k_i} D_{k_j} v_k = \phi \text{ whenever } n < m \text{ and } i > j. \tag{1}$$

Let $d_a := d_{M_a(S)}$ and choose sequences of points $\{c_{k_n}\}$, $\{e_{k_n}\}$ such that

$$c_{k_n} \in d_a(C_{k_n}) \quad \text{and} \quad e_{k_n} \in d_a(D_{k_n} v_k). \tag{2}$$

Let $E_k := \bigcup_{i=1}^{\infty} \bigcup_{i<j} C_{k_i} D_{k_j} v_k$ and $F_k := \bigcup_{j=1}^{\infty} \bigcup_{i>j} C_{k_i} D_{k_j} v_k$, and define the function h_k by $h_k := \chi_{E_k} - \chi_{F_k}$.

Noting that $h_k \in M_a(S)^*$ we now show that (in the norm of $M_a(S)^*$)

$$||h_k + g|| \geq 1 \quad \text{for all } g \in C(S). \tag{3}$$

If not, for some $g \in C(S)$ we can find $\varepsilon > 0$ such that

$$||h_k + g|| \leq 1-\varepsilon \quad .$$

In particular for $\nu_\alpha, \mu_\beta \in P(M_a(S))$ with $\text{supp}(\nu_\alpha) \subset C_{k_n}$ and $\text{supp}(\mu_\beta) \subset D_{k_m} v_k$, we have

$$|h_k(\nu_\alpha * \mu_\beta) + g(\nu_\alpha * \mu_\beta)| \leq 1-\varepsilon$$

and so (recalling our definition of h_k)

$$\begin{cases} \text{if } n < m \text{ then } |1 + g(\nu_\alpha * \mu_\beta)| \leq 1-\varepsilon \\ \text{if } n > m \text{ then } |-1 + g(\nu_\alpha * \mu_\beta)| \leq 1-\varepsilon . \end{cases}$$

Letting (ν_α) shrink to $\bar{c}_{k_n}$ and (μ_β) to $\bar{e}_{k_m}$ we thus get (by continuity of g)

$$\begin{cases} \text{if } n < m \text{ then } |1 + g(c_{k_n} e_{k_m})| \leq 1-\varepsilon \\ \text{if } n > m \text{ then } |-1 + g(c_{k_n} e_{k_m})| \leq 1-\varepsilon . \end{cases}$$

It follows that

$$G(c_{k_n}, e_{k_m}) = g(c_{k_n} e_{k_m}) \begin{cases} < -\varepsilon & \text{if } n < m \\ > \varepsilon & \text{if } n > m \end{cases}$$

and so $\{G(x,.) : x \in W\}$ is not relatively weakly compact, by Grothendieck's Theorem (-see Appendix B.7). This contradicts the observation at the beginning of our proof. By this conflict item (3) holds.

Since the V_k's are pairwise disjoint and $C_{k_n} D_{k_m} v_k \subset V_k$ for all $n,m,k \in \mathbb{N}$ we have that $\{t_k\} \to \sum_{k=1}^{\infty} t_k h_k + C(S)$ defines a linear mapping of ℓ^∞ into $M_a(S)^*/C(S)$. Noting that $\|h_k\| = 1$, item (3) implies that

$$\|\{t_k\}\|_\infty = \|\sum_{k=1}^{\infty} c_k h_k + C(S)\|_{M_a(S)^*/C(S)}$$

and so the mapping $\{t_k\} \to \sum_{k=1}^{\infty} t_k h_k + C(S)$ is isometric.

Similarly the mapping $\{t_k\} \to \sum_{k=1}^{\infty} t_k h_k + LWUC(S)$ of ℓ^∞ into $M_a(S)^*/LWUC(S)$ is linear and isometric.

THEOREM. *Let* S *be a nondiscrete and right cancellative foundation semigroup with an identity element. Then there exists a subspace* P *of* $M_a(S)^*$ *such that* P* *is a linear isometric copy of* $(\ell^\infty)^*$ *and the restriction of the radical of* $M_a(S)^{**}$ *to* P *is* P*. *In particular the radical of* $M_a(S)^{**}$ *is nonseparable.*

Proof. Let $A := \{\phi \in M_a(S)^{**} : \phi(f) = 0$ for all f in LWUC(S)$\}$. For all $\psi \in M_a(S)^{**}$, $\nu \in M_a(S)$ and $h \in M_a(S)^*$, we have that $\nu o h \in$ LWUC(S), by (the left handed version of) Lemma 4.3.2; consequently

$$h^{\phi}(\nu) := \phi(\nu o h) = 0 \quad \text{and so}$$

$$\phi o' \psi(h) := \psi(h^{\phi}) = 0 \quad (\phi \in A).$$

Thus A is a right ideal of $M_a(S)^{**}$ such that $A o' M_a(S)^{**}$ is zero. Hence $A \subseteq R_a(S)$, (see e.g. Rickart, "General theory of Banach algebras", Van Nostrand (1960); Theorem 2.3.5(ii)).

By the preceding Proposition, there exists an isometric linear map π of ℓ^{∞} into $M_a(S)^*/\text{LWUC}(S)$. So for some closed subsapce P of $M_a(S)^*$, we have $\pi(\ell^{\infty})$ dense in $P/\text{LWUC}(S)$. Hence the inverse map π^{-1} (defined on $\pi(\ell^{\infty})$) extends to a unique isometric linear map τ of $P/\text{LWUC}(S)$ onto ℓ^{∞}. The dual map $\tau^* : (\ell^{\infty})^* \to (P/\text{LWUC}(S))^*$ is isometric, linear and onto. But then $A = \text{LWUC}(S)^{\perp}$ can be identified with the dual of $M_a(S)^*/\text{LWUC}(S)$. So each element of $(P/\text{LWUC}(S))^*$ can be identified with the restriction of some element of A to P. This completes our proof (on noting that $(\ell^{\infty})^*$ is non-separable).

Let S be any topological semigroup satisfying the hypothesis of Theorem 2.11 and note that WUC(S)* is a Banach algebra under an Arens operation given by

$$\alpha \circledcirc' \beta(f) := \alpha(\beta o f) \quad \text{and} \quad \beta o f(x) := \beta({}_x f),$$

for all $\alpha, \beta \in$ WUC(S)*; $f \in$ WUC(S) and $x \in S$. Now, by Theorem 2.11, $\text{card}(\text{IM}(\text{WUC}(S))) \geq 2^c$. Let $B := \{\psi \in \text{WUC}(S)^* : \psi(1) = 0$ and $\psi({}_x f) = \psi(f)$, for all $x \in S$ and $f \in \text{WUC}(S)\}$. Note that B is a right ideal of WUC(S)* such that $B \circledcirc' B = \{0\}$. By the result referred to Rickart's book in the proof of the preceding Theorem, we have that the radical of the algebra WUC(S)* contains B. Now fix $\mu_o \in$ IM(WUC(S)) and note that IM(WUC(S)) $- \mu_o$ is contained in B, and so $\text{card}(B) \geq 2^c$. Consequently the radical of WUC(S)* is nonseparable.

PROBLEM. Let S be a cancellative infinite foundation semigroup. We conjecture that the radical of the algebra WUC(S)* is nonseparable. (The preceding remarks say that, to a large extent, the conjecture is true if there exists a left invariant mean on WUC(S) and S is not compact. Even for groups, this conjecture is still open [52].)

REFERENCES. Let G be a locally compact topological group and $\mathcal{R}$ the radical of $L^\infty(G)^*$. Civin and Yood [19] showed that $\mathcal{R}$ is infinite dimensional for G nondiscrete and abelian or G = Z. E.E. Granirer [52] proved that $\mathcal{R}$ is nonseparable if G is amenable or nondiscrete. Our theorem above is inspired by the latter paper of Granirer. In [56] Gulick showed that the quotient space $L^\infty(G)/C(G)$ is nonseparable if G is not extremely disconnected. Other results on the radical of $L^\infty(G)$ or UC(G)* can be found in [16].

5 Characterizations of absolutely continuous measures

This chapter forms a natural continuation of Chapter 2 bearing in mind our experiences in Chapter 3. The message of Chapter 2 is that one can study various spaces of measures, on a topological semigroup, that can be different in general but all coinciding with the group algebra in the case of a locally compact topological group; and further one such space, namely the space of measures which are absolutely continuous, seems to have many interesting properties. Chapter 3 then presents a very large class of topological semigroups, the so called stips, which have attractive topological "homogeneous" properties and include the class of all locally compact topological groups as a "very special" case.

In this chapter we shall study various characterizations of the absolutely continuous measures on a stip and also include some results very close to characterizing such measures.

Blanket Assumption. *Throughout this chapter* S *denotes a stip unless otherwise explicitly stated.*

1. CONTINUITY OF MEASURES UNDER TRANSLATION ON A STIP

Noting that the map $(x,\mu) \to \bar{x}*\mu$ of $S \times M(S)$ into $M(S)$ defines a left action of S on M(S), the following result follows as a special case of Theorem 3.4.6. (If S is also the foundation of $M_a(S)$, then one may also deduce our next Theorem from Corollary 2.2.4.)

1.1 THEOREM. *If* $\mu \in M(S)$, *the following items are equivalent:*

(i) *the map* ℓ_μ *is weakly continuous;*
(ii) *the map* ℓ_μ *is norm continuous.*

1.2 COROLLARY. *Let* $\mu \in M_a(S)$ *and* V *be a neighbourhood of* 1. *Then there exists a* δ*-isolated idempotent* e *in* V *such that*

$$supp(\mu) \subseteq eSe.$$

Proof. We may assume μ to be positive and find a σ-compact set M such that $||\mu|| = \mu(M)$. Recalling Lemma 2.1.4, our preceding Theorem says that $V_n := \{x \in S : ||\bar{x}*\mu - \mu|| < \frac{1}{n}\}$ is an open neighbourhood of 1 for all $n \in \mathbb{N}$. Since S_1 is dense in S (by Theorem 3.2.2), we can find a countable set $A := \{x_n \in V_n \cap S_1 : n \in \mathbb{N}\}$. By Lemma 3.3.4 we can find $e \in V \cap E_S^\delta$ such that $AM \subseteq eSe$. Now by definition of V_n we have

$$|\mu(S\setminus AM) - \bar{x}_n*\mu(S\setminus AM) \leq ||\mu-\bar{x}_n*\mu|| \to 0$$

and so $\mu(S\setminus AM) = \lim_n \bar{x}_n*\mu(S\setminus AM) = 0$ since each $\bar{x}_n*\mu$ is concentrated on AM. Hence μ is concentrated on AM. Since eSe is closed we conclude that $\mathrm{supp}(\mu) \subseteq \overline{AM} \subseteq eSe$.

1.3 REMARK. If S is a foundation semigroup then, with Exercise 3.3.7(i) in mind, one can easily note that

$$\cup\{\mathrm{supp}(\mu) : \mu \in M_a(S)\} = \chi_o - \mathrm{clo}(S_1).$$

Recalling Exercise 3.3.7(iv) we note the distinction between the foundation of $M_a(S)$ and the union of the supports of measures in $M_a(S)$. [The two notions coincide in the case of a group.]

For a general locally compact topological semigroup H, Exercise 2.1.15 (f) teaches us that we may have measures $\nu,\mu \in M(H)$ such that $\mu \in M_n(H)$ and $\nu << |\mu|$ but $\nu \notin M_n(H)$.

Contrary to this, for a stip S we have $M_n(S) = M_a(S)$ and so $M_n(S)$ is solid. Partly towards this end we have the following result.

1.4 LEMMA. *Let* $A_1(S) := \{\eta \in M(S) : \ell_\eta$ *is weakly continuous at* $1\}$, $\mu \in A_1(S)$ *and* $\nu \in M(S)$ *with* $\nu << |\mu|$. *Then* $\nu \in A_1(S)$.

Proof. Evidently $A_1(S)$ is a norm-closed linear subspace of M(S). Given $\varepsilon > 0$ and $f \in C_o(S)$, we note that

$$||f.(\bar{x}*\mu) - \bar{x}*(f.\mu)|| \leq |\mu| (|_xf-f|)$$

and therefore

$$U := \{x \in S : ||f.(\bar{x}*\mu) - \bar{x}*(f.\mu)|| < \varepsilon\}$$

is a neighbourhood of 1.

Since $\mu \in A_1(S)$, if $h \in M(S)^*$ with $||h|| = 1$ then

$$V := \{x \in S : |h(f.(\mu-\bar{x}*\mu))| < \varepsilon\}$$

is a neighbourhood of 1. So $U \cap V$ is a neighbourhood of 1. But

$$U \cap V \subseteq \{x \in S : |h(f.\mu) - h(\bar{x}*(f.\mu))| < 2\varepsilon\} .$$

Hence $f.\mu \in A_1(S)$. Since $\{f.\mu : f \in C_o(S)\}$ is dense in $L(S,|\mu|)$, it follows that $\nu \in A_1(S)$.

1.5 THEOREM. *If* $\mu \in M(S)$, *then the following items are equivalent:*

(i) ℓ_μ *is weakly continuous at the point* 1;

(ii) $\ell_{|\mu|}$ *is weakly continuous at the point* 1;

(iii) ℓ_μ *is weakly continuous on* S.

Proof. Evidently (iii) implies (i). The equivalence of (i) and (ii) is a special case of Lemma 1.4. So to complete the proof we assume item (i) and prove (iii). In view of Lemma 1.4 we may assume that μ is a positive measure.

Let C be a compact subset and (x_α) a net converging to x in S. There exists a subnet (y_β) of (x_α) such that

$$\lim_\beta \bar{y}_\beta*\mu(C) = \underline{\lim}_\alpha \bar{x}_\alpha*\mu(C). \tag{1}$$

By the Hahn-Banach theorem, one can find a real-valued functional $h \in M(S)^*$ such that, for all real-valued $\nu \in M(S)$, we have

$$- ||\nu|| \leq \underline{\lim}_\beta \bar{y}_\beta*\nu(C) \leq h(\nu) \leq \overline{\lim}_\beta \bar{y}_\beta*\nu(C) \leq ||\nu|| . \tag{2}$$

So, in particular

$$h(\mu) = \lim_\beta \bar{y}_\beta*\mu(C). \tag{3}$$

Let $\varepsilon > 0$ be given. Then item (i) implies that

$$U := \{u \in S : |\bar{u}*\mu(x^{-1}C)-\mu(x^{-1}C)| < \varepsilon\} \cap \{u \in S : |h(\bar{u}*\mu)-h(\mu)| < \varepsilon\}$$

is a neighbourhood of 1. Since S_1 is dense in S we can find $u \in int(U) \cap S_1$. Now $(xU)u^{-1}$ is a neighbourhood of x, by Theorem 3.2.6(i). So eventually $(y_\beta) \subseteq (xU)u^{-1}$.

Hence there is a net $(u_\beta) \subseteq U$ such that, eventually

$$y_\beta u = xu_\beta. \tag{4}$$

Now

$$\begin{aligned}
\bar{x}*\mu(C) &\leq \underline{\lim}\ \overline{xu_\beta}*\mu(C) + \varepsilon && \text{by definition of } U \\
&= \underline{\lim}_{\beta}\ \overline{y_\beta u}*\mu(C) + \varepsilon && \text{by item (4)} \\
&\leq h(\bar{u}*\mu) + \varepsilon && \text{by item (2)} \\
&\leq h(\mu) + 2\varepsilon && \text{by definition of } U \\
&= \lim_{\beta}\ \bar{y}_\beta*\mu(C) + 2\varepsilon && \text{by item (3)} \\
&= \underline{\lim}_{\alpha}\ \bar{x}_\alpha*\mu(C) + 2\varepsilon && \text{by item (1)} \\
&\leq \overline{\lim}_{\alpha}\ \bar{x}_\alpha*\mu(C) + 2\varepsilon \\
&\leq \bar{x}*\mu(C) + 2\varepsilon && \text{by Lemma 1.4.1(ii).}
\end{aligned}$$

Consequently $\mu \in M_a^\ell(S)$ and item (iii) follows, by Lemma 2.1.4.

1.6. *Summarising Corollary* 2.2.4(i), *Theorems* 1.1 *and* 1.5, *for each* $\mu \in M(S)$ *we have the following items equivalent*:

(i) $\mu \in M_a^\ell(S)$;

(ii) $\mu \in M_n^\ell(S)$;

(iii) $|\mu| \in M_n^\ell(S)$;

(iv) ℓ_μ *is weakly continuous at the point* 1.

1.7. Theorem 1.1 was first proved by Sleijpen [94] in a way relying heavily on measure theoretic techniques. The proof given here is due to Dzinotyiweyi [34]. The other results of this section are taken from Sleijpen ([94] and [92]).

2. MEASURES WITH SEPARABLE ORBITS

Recalling Definitions 3.4.1 we consider weakly and norm separable left orbits of measures in M(S) in terms of the left action of S on M(S) given by $(x,\mu) \to \bar{x}*\mu$ for all $x \in S$ and $\mu \in M(S)$.

First we have the following consequence of Theorem 3.4.4 and item 1.6.

2.1 THEOREM. *Let* U *be a compact neighbourhood of* 1 *and* $\mu \in M(S)$.

(i) *If* μ *has a weakly separable left orbit over* U, *then* $\bar{x}*\mu \in M_a^{\ell}(S)$ *for all* $x \in S_1$;

(ii) *If* $\mu \in M_a^{\ell}(S)$, *then* μ *has a norm separable left orbit over* U.

Proof. Let $\varepsilon > 0$ be given, $y \in S$ fixed and consider the set

$$O(y) := \{s \in S : \|\bar{s}*\mu - \bar{y}*\mu\| \leq \varepsilon\} .$$

To prove item (i) it is sufficient to show that O(y) is closed and Theorem 3.4.4 will imply the result. To the latter end we first note that the continuity of the function $s \to \bar{s}*\mu(f)$ defined on S ($f \in C(S)$) implies that

$$O_f(y) := \{s \in S : |\bar{s}*\mu(f) - \bar{y}*\mu(f)| \leq \varepsilon\}$$

is closed in S. Since M(S) is the first continuous dual of $C_o(S)$, it is trivial to note that, from

$$\|\bar{s}*\mu - \bar{y}*\mu\| = \sup\{|\bar{s}*\mu(f) - \bar{y}*\mu(f)| : f \in C_o(S), \|f\|_S \leq 1\},$$

we have that

$$O(y) = \cap\{O_f(y) : f \in C_o(S), \|f\| \leq 1\} .$$

Thus O(y) is closed and item (i) follows.

Now item 1.6 says that $\mu \in M_a^{\ell}(S)$ implies that $\{\bar{x}*\mu : x \in U\}$ is norm compact and so item (ii) follows.

2.2 Notes. The subject of absolute continuity of measures with separable orbits particularly on locally compact groups, has been studied quite extensively. This seems to have started with a paper of R. Larsen [70], and subsequently various other papers appeared - see e.g. Tam [100]; T.S. Liu, A. van Rooij and J.K. Wang [73], G.L.G. Sleijpen [93] and H.A.M. Dzinotyiweyi ([33] and [34]).

In this chapter, Theorem 2.1(i), (or rather its general form: Theorem 3.4.4) has enormous applications. In fact almost every Theorem we prove in this chapter can be traced back to have its roots in Theorem 2.1(i).

Theorem 2.1(i) was first proved for stips admitting a certain measure theoretic condition in [93] and its proof in the general form given above first appeared in [33].

2.3 Remark. In general there are stips (even foundation semigroups with identity element) admitting measures having norm separable left orbits but are not absolutely continuous. See for instance Example 4.4.

In preparation for our next Theorem we give a characterization of $M_a^{\ell}(S)$ in the following Lemma.

2.4 LEMMA. *The following items are equivalent for any* $\mu \in M(S)$:

(i) $\mu \in M_a^{\ell}(S)$;

(ii) *for each* δ*-isolated idempotent* e *of* S, *we have*

$$\bar{e}*\mu*\bar{e} \in M_a(eSe).$$

Proof. Evidently (i) implies (ii). We now assume (ii) and prove (i). Let U be an open relatively compact neighbourhood in S and A any countable subset of $\bar{U} \cap S_1$. By (the right handed form of) Lemma 3.3.4 we can find a δ-isolated idempotent $e \in S$ such that

$$A(\text{supp}|\mu| \cup \{1\}) \subseteq eSe \quad \text{and hence}$$

$$\bar{a}*\mu = \bar{a}*\bar{e}*\mu*\bar{e} \quad \text{for all } a \in A, \text{ and } A \subseteq eSe \cap \bar{U} .$$

So item (ii) implies that $\{\bar{a}*\bar{e}*\mu*\bar{e} : a \in A\}$ is relatively norm compact, by the compactness of $\bar{U} \cap eSe$. Thus $\{\bar{a}*\mu : a \in A\}$ is relatively norm compact for all countable subsets A of $\bar{U} \cap S_1$. Hence, we have that

$$\{\bar{x}*\mu : x \in \bar{U} \cap S_1\} \text{ is relatively norm compact.}$$

Since each norm-compact subset of $M(S)$ is weak*-closed and each weak*-closed subset of $M(S)$ is norm-closed, the fact that the weak*-closure of $\{\bar{x}*\mu : x \in \bar{U} \cap S_1\}$ is $\{\bar{x}*\mu : x \in \bar{U}\}$ implies that the latter set is norm compact. Hence $\ell_{\mu,\bar{U}}$ is norm continuous, by Lemma 2.2.1 and item (i) follows, by item 1.6.

2.5 DEFINITIONS. Recalling the definition of equi-absolute continuity just before Lemma 2.1.6, we now formally define spaces of such measures. Let

$$M^{\ell}_{eq-a}(S) := \{\mu \in M(S) : \bar{x}*|\mu| << |\nu|*|\mu| \quad \text{for all } x \in \text{supp}(\nu) \text{ and } \nu \in M(S)\}$$

$$M^{r}_{eq-a}(S) := \{\mu \in M(S) := |\mu|*\bar{x} << |\nu|*|\mu| \text{ for all } x \in \text{supp}(\nu) \text{ and } \nu \in M(S)\}$$

$$M_{eq-a}(S) := M^{\ell}_{eq-a}(S) \cap M^{r}_{eq-a} .$$

Here S is any C-distinguished topological semigroup.

It would be interesting to investigate the properties of the space of equi-absolutely continuous measures. In the following example we show that the space may differ from that of absolutely continuous measures for a general locally compact topological semigroup.

2.6 EXAMPLE. Let $G_2 := \prod_{i=1}^{\infty} Z_3$ be the product topological group of countably many copies of the (discrete) cyclic group of 3 elements $Z_3 := \{0,1,2\}$. Let $H := G_3 \cup \{\theta\} \cup \{(n,x) \in \mathbb{N} \times G_3$: the set $\{m \in \mathbb{N} : x(m) \neq 0\}$ has at most n elements$\}$ with the operation

$$y(n,x) = (n,x)y := xy, \ ab=z\theta=\theta z:= \theta \quad (y \in G_3;\ a,b,(n,x) \in \mathbb{N}\times G_3 \text{ and } z \in H)$$

Then, with θ isolated, H is a locally compact topological semigroup. The product measure $\mu := \prod_{i=1}^{\infty} (\frac{1}{2}\bar{0} + \frac{1}{2}\bar{1})$ defined on G_3 (and hence on H) can be shown to be equi-absolutely continuous but not absolutely continuous.

For a stip, we show in our next Theorem that equi-absolutely continuous measures are precisely the absolutely continuous measures. Towards this end we first show that $M^{\ell}_{eq-a}(S)$ is solid.

2.7 LEMMA. *Let* $\mu \in M^{\ell}_{eq-a}(S)$ *and* $\eta \in M(S)$ *with* $\eta << |\mu|$. *Then* $\eta \in M^{\ell}_{eq-a}(S)$.

Proof. Let $\nu \in M(S)$ be fixed. Since $|\eta| << |\mu|$, we evidently have

$$|\nu|*|\eta| << |\nu|*|\mu| \quad . \tag{1}$$

By the Hahn-decomposition Theorem, we can find non-negative measures ρ_1 and ρ_2 in $M(S)$ such that

$$|\nu|*|\mu| = \rho_1 + \rho_2, \quad \rho_1 << |\nu|*|\eta| \quad \text{and} \quad \rho_2 \perp |\nu|*\eta.$$

Recalling (1), we see that

$$\rho_1 << |\nu|*|\eta| \quad \text{and} \quad |\nu|*|\eta| << \rho_1. \tag{2}$$

We can find a Borel subset F of S such that ρ_1 is concentrated on F and $\rho_2(F) = 0$.

Now suppose that, for some Borel $E \subset S$, we have $|\nu|*|\eta|(E) = 0$. Then

$$|\nu|*|\mu|(E \cap F) = \rho_1(E) = 0.$$

So

$$|\mu|(x^{-1}(E \cap F)) = \bar{x}*|\mu|(E \cap F) = 0 \text{ for all } x \in \text{supp}(\nu),$$

since $\mu \in M^{\ell}_{eq-a}(S)$.

Recalling that $|\eta| << |\mu|$ we get

$$\bar{x}*|\eta|(E \cap F) = 0 \quad \text{for all } x \in \text{supp}(\nu). \tag{3}$$

Now clearly $\text{supp}(\bar{x}*|\eta|) \subseteq \text{supp}(|\nu|*|\eta|)$, for all $x \in \text{supp}(\nu)$. Hence, since $|\nu|*|\eta|$ is concentrated on F (by (2)), we have that $\bar{x}*|\eta|$ is concentrated on F, for all $x \in \text{supp}(\nu)$. This together with (3) says that $\bar{x}*|\eta|(E) = 0$ and hence that

$$\bar{x}*|\eta| << |\nu|*|\eta| \quad \text{for all } x \in \text{supp}(\nu).$$

This completes our proof.

2.8 THEOREM. *If* $\mu \in M(S)$ *then the following items are equivalent:*

(i) $\mu \in M^{\ell}_{eq-a}(S)$;

(ii) $\mu \in M^{\ell}_{a}(S)$.

Proof. That (ii) implies (i) follows from Lemma 2.1.6. We now assume (i) and prove (ii). In view of Lemma 2.7 and the fact that $M^{\ell}_{a}(S)$ is now a closed solid space, we assume that μ is a positive measure with compact support K (say).

First suppose that 1 is δ-isolated. Let V be a compact neighbourhood of 1. Proposition 3.3.8 says that there exists a countable subset B of $A := VK \cup V$ and a G_δ compact subgroup G contained in $F := \mathrm{int}(V)$, such that items (i) to (iii) of the Proposition hold.

Let $\{b_k\}$ be an enumeration of $B \cap GV$ and π be the normalised Haar measure on G. By Proposition 3.3.8(ii) we have that

$$\pi * \bar{x} * \bar{g} = \pi * \bar{x}, \quad \bar{x} * \pi = \bar{g} * \bar{x} * \pi \quad (g \in G \text{ and } x \in A).$$

Hence

$$\pi * \bar{x} * \pi = \bar{x} * \pi = \pi * \bar{x} \quad (x \in A). \tag{1}$$

Let $\nu := \sum_{k=1}^{\infty} 2^{-k} \pi * \bar{b}_k$ and note that $\mathrm{supp}(\nu) = \overline{GB} = GV$, by Proposition 3.3.8 (i). Since $\mu \in M^{\ell}_{eq-a}(S)$, we thus have that

$$\bar{x} * \mu \ll \nu * \mu \quad \text{for all } x \in GV.$$

Consequently

$$\bar{x} * \pi * \mu \ll \nu * \mu \text{ for all } x \in GV. \tag{2}$$

Now Proposition 3.3.8(iii) and the fact that C(S) is dense in $L^1(\nu * \mu)$, imply that $\{f \in L^1(\nu * \mu) : \bar{x} * \pi(f) = f(x) \text{ for } \nu * \mu\text{-almost all } x \in GA\}$ is a separable subset of $L^1(\nu * \mu)$. Hence, by (1) and (2), we have that $\{\bar{x} * \pi * \mu : x \in GV\}$ is separable.

By Theorem 2.1(i), we have that

$$\bar{x} * \pi * \mu \in M^{\ell}_{a}(S) \quad \text{for all } x \in S_1. \tag{3}$$

Since 1 is δ-isolated, Theorem 3.3.6 implies the existence of a countable subset $C := \{c_n : n \in \mathbb{N}\}$ of S_1 with $1 \in \bar{C}$. So $1 \in \overline{CG}$.

Taking $\nu_o := \sum_{n=1}^{\infty} 2^{-n} \bar{c}_n * \pi$ we have $1 \in \overline{CG} = \text{supp}(\nu_o)$ and so

$$\mu << \nu_o * \mu.$$

By (3) we have $\nu_o * \mu \in M_a^{\ell}(S)$ and hence $\mu \in M_a^{\ell}(S)$.

Now suppose 1 is not δ-isolated. It is trivial to note that $\mu \in M_{eq-a}^{\ell}(S)$ implies that $\bar{e} * \mu * \bar{e} \in M_{eq-a}(eSe)$ for all $e \in E_S$. By the above argument and Lemma 2.4 we thus have

$$\mu \in M_a(S).$$

This completes our proof.

The proof of the preceding Theorem is taken from Sleijpen [98]. The Theorem is essentially a consequence of Theorem 2.1(i) which uses the notion of measure with separable orbits. It remains an open problem whether one can supply a proof independent of the notion of separable orbits. Example 2.6 is taken from [96].

3. MEASURES VANISHING ON EMACIATED SETS

3.1 DEFINITIONS. A subset A of S is said to be *left-emaciated* if AA^{-1} is not a neighbourhood of the identity element of S. The collection of all compact G_δ left-emaciated subsets of S is denoted by L_e and we define $EM^{\ell}(S) := \{\mu \in M(S) : \mu(F) = 0 \text{ for all } F \in L_e\}$.

We denote the collection of all $M_a(S)$-*negligible subsets* of S by N_a.

3.2 PROPOSITION. *Each compact left-emaciated subset,* F, *of* S *is both meagre and* $M_a(S)$-*negligible.*

Proof. That F is $M_a(S)$-negligible follows from Lemma 2.1.9 and our definitions.

To see that F is meagre, let $B := S \setminus FF^{-1}$ and note that $1 \in \bar{B}$ and $BF \cap F = \phi$. If $x \in F$ then $x \in \overline{Bx}$ and $Bx \cap F = \phi$, which shows that $\text{int}(F) = \phi$. Since F is closed, we have that F is meagre.

The proof for the following Proposition is easy and left as an exercise for the reader.

3.3 PROPOSITION. *For all* $\nu \in M(S)$ *and* $\mu \in EM^{\ell}(S)$ *with* $\nu << |\mu|$, *we have that* $\nu \in EM^{\ell}(S)$. *Further* $EM^{\ell}(S)$ *is norm closed in* $M(S)$.

We now begin an argument culminating in the result that every measure in $EM^{\ell}(S)$ is left equi-absolutely continuous. We partition part of our proof into some lemmas.

3.4 LEMMA. *Let* $\mu \in M(S)^{+}$, $B \subseteq S$ *and* K *a compact subset of* S. *Then for each* $x \in \bar{B}$, $v \in S$ *and compact* $F \subset S$ *with* $F \subseteq x^{-1}K \setminus B^{-1}K$ *the set* $\{t \in S : \bar{v}*\bar{t}*\mu(F) \neq 0\}$ *is meagre.*

Proof. There exists a net (x_α) in B converging to x. For each $t \in S_1$ and compact neighbourhood T of t, we have that $((xvT)t^{-1})v^{-1} = (xvT)(vt)^{-1}$ is a neighbourhood of x, by Theorem 3.2.6(i). So there is a net (t_α) in T converging to t such that

$$x_\alpha vt = xvt_\alpha \quad \text{for all } \alpha.$$

Now since $F := x^{-1}K \setminus B^{-1}K$ an easy manipulation shows that

$$(vt)^{-1}F \cap (vt_\alpha)^{-1}F = \phi \quad \text{for all } \alpha. \tag{1}$$

Let $W_n := \{s \in S : \bar{v}*\bar{s}*\mu(F) \geq \frac{1}{n}\}$, for all $n \in \mathbb{N}$. By the upper semi-continuity of the map $s \to \bar{s}*\mu(v^{-1}F)$ (recall Lemma 1.4.1), we have that W_n is closed.

From (1) and Lemma 1.4.1(ii), we have that

$$\begin{aligned}\lim_\alpha \bar{v}*\bar{t}_\alpha*\mu(F) &= \lim_\alpha \mu((vt_\alpha)^{-1}F \setminus (vt)^{-1}F) \\ &= \lim_\alpha \mu|_{S\setminus (vt)^{-1}F}((vt_\alpha)^{-1}F) \\ &\leq \mu|_{S\setminus(vt)^{-1}F}((vt)^{-1}F) = 0.\end{aligned}$$

Since t was arbitrarily chosen in S_1, S_1 is dense in S and W_n is closed, it follows that W_n is nowhere dense.

Consequently $\{t \in S : \bar{v}*\bar{t}*\mu(F) \neq 0\} = \bigcup_{n=1}^{\infty} W_n$ is meagre.

3.5. In this item we develop an argument that is needed in the following two Lemmas.

Let K,F and V_o be compact subsets of S such that K is a G_δ-subset and V_o is a neighbourhood of 1. Thus there is a decreasing sequence $\{U_n\}$ of open subsets such that $K = \bigcap_{n=1}^{\infty} U_n$.

Remembering Theorem 3.2.6, we can find a sequence $\{V_n\}$ of open neighbourhoods of 1 and a sequence $\{v_n\}$ of elements in S_1 such that

$$V_1 \subseteq V_o,\ v_n \in V_n$$

$$\bar{V}^2_{n+1} \subseteq V_n^{-1} v_n \cap V_n$$

$$V_n^{-1}(V_n K) \cap F \subseteq U_n; \quad \text{for all } n \in \mathbb{N}.$$

Let $H := \bigcap_{n=1}^{\infty} V_n$. Then one can easily note that H is a compact G_δ-subsemigroup of S with identity element 1. We then get

$$H^{-1}(HK) \cap F = \bigcap_{n=1}^{\infty} (V_n^{-1}(V_n K) \cap F) \subseteq \bigcap_{n=1}^{\infty} U_n = K$$

and hence

$$H^{-1}(HK) \cap F = K \cap F \tag{1}$$

Similarly taking V_oF in place of F in (1) we get

$$H^{-1}(HK) \cap V_oF = K \cap V_oF \tag{2}$$

Let e be an idempotent in the minimal ideal of H. Then from Theorem 1.2.3, we have that G := eHe is a compact subgroup of S with identity element e. For all h ∈ H, we have that

$$\begin{aligned} H \cap F &\subseteq h^{-1}(H^{-1}(HK) \cap V_oF) \cap F \\ &\subseteq h^{-1}K \cap F && \text{by (2)} \\ &\subseteq H^{-1}(HK) \cap F = K \cap F && \text{by (1).} \end{aligned}$$

Consequently

$$K \cap F = h^{-1}K \cap F \quad \text{for all } h \in H. \tag{3}$$

3.6 LEMMA. *Let* K *be a compact* G_δ*-subset of* S *and* $\mu \in M(S)^+$ *with* $M := supp(\mu)$ *compact. Let* $x \in clo\{y \in S : \bar{y}*\mu(K) = 0\}$. *Then there exists a* $B \subseteq S$ *and an* $e \in S$ *such that*

(i) $B^{-1}K$ *is* μ*-measurable and* $\bar{\mu}(B^{-1}K) = 0$;

(ii) $ex \in \bar{B}$;

(iii) $\overline{ex}*\mu(K) = \bar{x}*\mu(K)$.

Proof. Let X be a compact neighbourhood of x and take $F := XM$ in the argument presented in item 3.5. Assuming the notation of 3.5, let $P := \{y \in eX : \bar{y}*\mu(K) = 0\}$. By equation (3) of item 3.5, we have $ex \in \bar{P}$. Hence we can choose $a_n \in V_n^{-1}(V_n ex) \cap P$, for all $n \in \mathbb{N}$. Let $B_1 := \{a_n : n \in \mathbb{N}\}$ and $B := H^{-1}(HB_1)$. We then have $B^{-1}K \subseteq B_1^{-1}H^{-1}(HK)$ and the latter set is clearly an F_σ-set. Now

$$\begin{aligned}\mu(B_1^{-1}H^{-1}(HK)) &\leq \sum_{n=1}^{\infty} \bar{a}_n*\mu(H^{-1}(HK)) = \sum_{n=1}^{\infty} \bar{a}_n*\mu(H^{-1}(HK) \cap V_oF) \\ &= \sum_{n=1}^{\infty} \bar{a}_n*\mu(K \cap V_oF) \quad \text{by 3.5} \qquad (2) \\ &= 0.\end{aligned}$$

Thus $B^{-1}K$ is contained in a set of μ-measure zero, and so $B^{-1}K$ is μ-measurable and $\bar{\mu}(B^{-1}K) = 0$. This proves (i).

To prove (ii), let U be an open neighbourhood of ex. Recalling the notation of 3.5, we claim that there exists an $n \in \mathbb{N}$ such that

$$(e\bar{V}_n e)^{-1}\,(e\bar{V}_n ex) \cap eX \subseteq GU. \qquad (1)$$

To verify this claim we first note that

$$\bigcap_{n=1}^{\infty} (e\bar{V}_n e)^{-1}\,(e\bar{V}_n(ex)) \cap eX \subseteq G^{-1}(Gex) \cap eX \subseteq e^{-1}Gx.$$

So

$$\bigcap_{n=1}^{\infty} (((e\bar{V}_n e)^{-1}\,(e\bar{V}_n ex) \cap eX)\backslash GU) = \phi.$$

Since G is a group GU is open, and so the above represents an intersection of closed subsets of the compact space eX. Hence for some fixed $n \in \mathbb{N}$ we must have

$$\bigcap_{k=1}^{n} (((e\bar{V}_k e)^{-1} (e\bar{V}_k ex) \cap eX) \setminus GU) = \phi.$$

Consequently item (1) follows, since $\{V_n\}$ is a contracting sequence. Now an easy exercise on our definitions shows that

$$V_n^{-1}(V_n ex) \cap eX \subseteq (eV_n e)^{-1}(eV_n ex) \cap eX. \qquad (2)$$

From (1) and (2) we have that $B_1 \cap FU \neq \phi$. So $B_1 \cap H^{-1}(HU) \neq \phi$ or $B \cap U \neq \phi$. It follows that $ex \in \bar{B}$ and (ii) is proved.

Finally, from equation (3) of 3.5 item (iii) follows.

3.7 LEMMA. *Let* K *be a compact* G_δ*-subset of* S *and* $\mu \in M(S)^+$ *such that* $M := supp(\mu)$ *is compact and* $\{y \in S : \overline{vy}*\mu(K) \neq 0\}$ *is meagre, for each* $v \in S$. *Then there exists a* $C \subseteq S$ *such that*

(i) $C^{-1}K$ *is* μ*-measurable and* $\bar{\mu}(C^{-1}K) = 0$;

(ii) $1 \in \bar{C}$.

Proof. Let U be a compact neighbourhood of 1 and take $F := UM$ in the argument presented in item 3.5. Assuming the notation of 3.5, let A_n be a finite subset of $U \cap \{y \in S : \overline{v_{n+1}} y*\mu(K) = 0\}$ such that

$$U \subset V_{n+2}^{-1}(V_{n+2}A_n) \quad \text{for all } n \in \mathbb{N}.$$

Let $C_1 := \bigcup_{n=1}^{\infty} v_{n+1}A_n$ and $C := H^{-1}(HC_1)$. Then arguing as in the proof of Lemma 3.6(i) one easily obtains our item (i).

To prove (ii), let O be any open neighbourhood of 1 with $O \subseteq U$. By Theorem 3.2.6(i) we can find $u \in O \cap S_1$ such that $(HO)u^{-1}$ is a neighbourhood of H and therefore we can find $n \in \mathbb{N}$ with $V_n \subseteq (HO)u^{-1}$. Since $u \in U \subset V_{n+2}^{-1}(V_{n+2}A_n)$, there exists an $a \in A_n$ such that

$$a \in V_{n+2}^{-1}(V_{n+2}u) \subseteq (V_{n+1}^{-1}v_{n+1})^{-1}(V_{n+2}u) \subseteq v_{n+1}^{-1}(V_{n+1}V_{n+2}u)$$

$$\subseteq v_{n+1}^{-1}(V_n u) \subseteq v_{n+1}^{-1}(HO).$$

Hence $v_{n+1}a \in HO$, so $C_1 \cap HO \neq \phi$. Consequently $C \cap O \neq \phi$, and so $1 \in \bar{C}$.

Now we can prove our first theorem in this section.

3.8 THEOREM. *We have that*

$$EM^{\ell}(S) \subseteq M^{\ell}_{eq-a}(S).$$

Proof. Let $\mu \in EM^{\ell}(S)$. To show that $\mu \in M^{\ell}_{eq-a}(S)$, in view of Proposition 3.3, we shall assume that μ is positive and $M := \mathrm{supp}(\mu)$ is compact. Let $\nu \in M(S)$ and K be a compact G_δ-subset of S such that $|\nu|*\mu(K) = 0$. Thus $0 = \int \bar{y}*\mu(K)d|\nu|(y)$, so $\bar{y}*\mu(K) = 0$ for ν-almost every $y \in \mathrm{supp}(\nu)$. Fix $x \in \mathrm{supp}(\nu)$ and note that $x \in \mathrm{clo}\{y \in S : \bar{y}*\mu(K) = 0\}$.

Now let B and e be chosen as in Lemma 3.6. We therefore have that $0 = \bar{x}*\mu(K) = \bar{\mu}((ex)^{-1}K\backslash B^{-1}K)$ if and only if $\mu(F) = 0$ for every compact G_δ-set F contained in $(ex)^{-1}K\backslash B^{-1}K$.

Let F be a compact G_δ-set with $F \subset (ex)^{-1}K\backslash B^{-1}K$. Then Lemma 3.4 says that $\{t \in S : \bar{v}*\bar{t}*\mu(F) \neq 0\}$ is meagre, for each $v \in S$.

Next Lemma 3.7 says there exists a $C \subseteq S$ such that

$$1 \in \bar{C},\ C^{-1}F \text{ is } \mu\text{-measurable and } \bar{\mu}(C^{-1}F) = 0.$$

Hence $F\backslash C^{-1}F$ is μ-measurable and $0 = \mu(F) = \bar{\mu}(F\backslash C^{-1}F)$ if and only if $\mu(A) = 0$ for each compact G_δ-set A for which $A \subseteq F\backslash C^{-1}F$. But that $A \subset F\backslash C^{-1}F$ implies that $AA^{-1} \cap C = \phi$. Since $1 \in \bar{C}$ we thus have $A \in L_e$ and so $\mu(A) = 0$ by our definition of $EM^{\ell}(S)$.

Hence $\bar{x}*\mu(K) = 0$ and so $\bar{x}*\mu << |\nu|*\mu$, for all $x \in \mathrm{supp}(\nu)$. Thus $\mu \in M^{\ell}_{eq-a}(S)$.

Combining Proposition 3.2 with Theorems 3.8 and 2.8 we have the following theorem.

3.9 THEOREM. *For every* $\mu \in M(S)$, *the following items are equivalent*:

(i) $\mu \in M_a^\ell(S)$;

(ii) $\mu \in M_{eq-a}^\ell(S)$;

(iii) $\mu \in EM^\ell(S)$.

As a consequence we now see that if a measure is singular with respect to every measure in $M_a(S)$, then such a measure is concentrated on an $M_a(S)$-negligible subset of S. (This is implied by Lemma 2.1.9 and Corollary 3.10).

3.10 COROLLARY. *Let* $\mu \in M(S)$ *be such* $\mu \perp M_a^\ell(S)$. *Then there exists a* σ*-compact subset* A *of* S *such that*

$$1 \notin int(AA^{-1}) \quad and \quad |\mu|(A) = \|\mu\| .$$

Proof. Without lossing generality we may assume that μ is positive and $M := supp(\mu)$ is compact. There exists a sequence $\{K_n\}$ in L_e such that

$$\mu(\bigcup_{n=1}^{\infty} K_n) = supp\{\mu(\bigcup_{n=1}^{m} F_n) : F_n \in L_e \text{ and } n \in \mathbb{N}\}. \tag{1}$$

Setting $K := \bigcup_{n=1}^{\infty} K_n$, we see from (1) that $\mu-\mu_{|K} \in EM^\ell(S)$ and hence that $\mu-\mu_{|K} \in M_a^\ell(S)$, by Theorem 3.9. It follows that $\mu = \mu_{|K}$ since $\mu \perp M_a^\ell(S)$ and $\mu_{|K} << \mu$. Hence

$$\|\mu\| = \mu(K).$$

From Lemma 3.4, for each $v \in S$, we have that

$$\{t \in S : \bar{v}*\bar{t}*\mu(K) \neq 0\} = \bigcup_{n=1}^{\infty} \{t \in S : \bar{v}*\bar{t}*\mu(K_n) \neq 0\}$$

is meagre. Hence, by Lemma 3.7, there exists a subset C such that $1 \in \bar{C}$, $C^{-1}K$ is μ-measurable and $\bar{\mu}(C^{-1}K) = 0$.

We can then find a σ-compact subset A of $D := K \setminus C^{-1}K$ such that

$$\mu(A) = \bar{\mu}(D) = \mu(K) = \|\mu\| .$$

That $A \subseteq K \setminus C^{-1}K$ implies that $AA^{-1} \cap C = \phi$. Since $1 \in \bar{C}$, it follows that $1 \notin \mathrm{int}(AA^{-1})$. This completes our proof.

3.11 Note. In the case where S is a locally compact group, the result that $EM^{\ell}(S) = M_a^{\ell}(S)$ was proved by S.M. Simons [91]. The extension to the case of stips is due to G.L.G. Sleijpen. The results given in this section are taken from Sleijpen [93].

It is trivial to note that in the case where S is a foundation semigroup, Borel emaciated sets are $M_a(S)$-negligible. In that case, a measure $\mu \in M(S)$ is in $M_a(S)$ as soon as $\mu(F) = 0$ for every compact $M_a(S)$-negligible subset F of S.

4. LOCALLY QUASI-INVARIANT MEASURES

We now resume studies introduced in Section 4 of Chapter 2. The results of this section are taken from [93] and [92].

4.1 DEFINITION. A measure $m \in M(S)$ is said to be *left locally quasi-invariant* if there exists a neighbourhood U of 1 such that

$$\bar{x}*m << |m| \quad \text{for all} \quad x \in U.$$

Recalling item 2.4.5, the reader can easily note that the notion of local quasi-invariance is different from that of equi-quasi-invariance introduced in Chapter 2. However the technique employed in the proof of Theorem 2.4.2 is similar to that employed in the following Theorem.

4.2 THEOREM. *We have that*

$$M_a(S) = \cup\{L(S,m) : m \in M_a(S)^+ \text{ and m is left locally quasi-invariant}\}.$$

Proof. Let U be a compact neighbourhood of 1 and $\mu \in M_a(S)$. It is sufficient to prove the existence of an $m \in M_a(S)^+$ such that $\mu << m$ and $\bar{x}*m << m$ $(x \in U)$. By Lemma 2.1.4, we have that $\{\bar{x}*|\mu| : x \in U^p\}$ is weakly compact $(p \in \mathbb{N})$. By Appendix B.4(iii), we can find a sequence $\{x_{p,n}\}_{n \in \mathbb{N}}$ such that $\{\bar{x}_{p,n}*|\mu| : n \in \mathbb{N}\}$ is weakly dense in $\{\bar{x}*|\mu| : x \in U^p\}$. Let

$$m := |\mu| + \sum_{1\leq p,n<\infty} 2^{-(n+p)}\, \bar{x}_{p,n} * |\mu|$$

and note that $\mu << m$ and $m \in M_a(S)^+$.

Now if $m(E) = 0$ for any Borel $E \subseteq S$ we have $\bar{x}_{p,n} * |\mu|(E) = 0$ $(n,p \in \mathbb{N})$ and hence $\bar{x} * |\mu|(E) = 0$ $(x \in \bigcup_{p=1}^{\infty} U^p)$. Noting that $xx_{p,n} \in \bigcup_{p=1}^{\infty} U^p$ for all $p,n \in \mathbb{N}$ and $x \in U$ we conclude that $\bar{x}*m << m$ $(x \in U)$.

It is nice to note that, using the notion of measures vanishing on emaciated sets, we have that left locally quasi-invariant measures are "very close" to being left absolutely continuous.

4.3 THEOREM. *Let* $m \in M(S)$ *be left locally quasi-invariant. Then* $\bar{x}*m \in M_a^{\ell}(S)$ *for all* $x \in S_1$.

Proof. There exists a neighbourhood U of 1 such that $\bar{y}*m << |m|$ $(y \in U)$. Let $x \in S_1$ be fixed. Since $S_1^{-1}x$ is a neighbourhood of 1 (by Theorem 3.2.6 (iii)), we can find a $y \in U \cap S_1$ and a $v \in S_1$ with $x = vy$.

Let $K \in L_e$. By noting that $K \cap (S\backslash KK^{-1})^{-1}K = \phi$ and $1 \in \mathrm{clo}(S\backslash KK^{-1})$, Lemma 3.4 implies that $\{t \in S : \bar{t}*|m|(K) > 0\}$ is meagre. Since yU^{-1} is a neighbourhood of 1, we can find a $t \in yU^{-1}$ such that $\bar{t}*|m|(K) = 0$. Choose $u \in U$ with $tu = y$. Then $\bar{y}*m(K) = \bar{u}*m(t^{-1}K) = 0$ and hence $\bar{y}*m \in EM^{\ell}(S)$.

By Theorem 3.9 we have $\bar{y}*m \in M_a^{\ell}(S)$. Consequently $\bar{x}*m = \bar{v}*\bar{y}*m$ is in $M_a^{\ell}(S)$.

In Theorem 4.3 one cannot replace S_1 by S.

4.4 EXAMPLE. Consider $S := \{0\} \cup \{\frac{1}{n} : n \in \mathbb{N}\}$ with maximum operation and the usual topology. Thus S is a foundation semigroup with identity element. Let $m := \bar{0} + \sum_{n=1}^{\infty} 2^{-n}\,(\overline{\tfrac{1}{n}})$ and note that $\bar{x}*m << m$ $(x \in S)$. However since $\bar{0} \notin M_a^{\ell}(S)$ and $\bar{0} << m$ we have that $m \notin M_a^{\ell}(S)$.

5. CONTINUITY OVER DIRECT TRANSLATES OF COMPACT SETS

Our definition of absolute continuity involves inverse translates of compacts, that is sets of the form $x^{-1}K$ and Kx^{-1} where $K \subseteq S$ is compact and $x \in S$. In this section we wish to involve direct translates of compact sets, that is sets of the form xK or Kx.

A.C. Baker and J.W. Baker [4] studied the collection of measures $\mu \in M(S)$ such that $x \to |\mu|(xK)$ and $x \to |\mu|(Kx)$ are maps of S into $\mathbb{R}$ continuous on S or at a given point of S; where S is any locally compact topological semigroup. Such spaces don't seem to be as nicely behaved as $M_a(S)$, in general. For a stip we however have the following characterization of one such space as $M_a^{\ell}(S)$.

5.1 THEOREM. *For every* $\mu \in M(S)$, *the following items are equivalent:*

(i) *For every compact* $C \subseteq S$ *the function* $x \to \mu(xC)$ *of* S *into* $\mathbb{C}$ *is continuous at* 1;

(ii) $\mu \in M_a^{\ell}(S)$.

Proof. To show that (i) implies (ii) let $K \in L_e$ be fixed and $\varepsilon > 0$ be given. There exists a compact neighbourhood U of 1 such that $|\mu|(UK\setminus K) < \varepsilon$. Taking $B := S\setminus KK^{-1}$ we note that $1 \in \bar{B}$ and $bK \cap K = \phi$ for all $b \in B$. So for all $x \in U \cap B$ we have that

$$|\mu(xK)| = |\mu(xK\setminus K)| \leq |\mu|(UK\setminus K) < \varepsilon ,$$

and hence, by (i), we get $\mu(K) = 0$ and $\mu \in EM^{\ell}(S)$. By Theorem 3.9 we have (ii).

Next we assume (ii) and prove (i). Since $M_a^{\ell}(S)$ is solid we may assume that μ is a positive measure. (Our proof is similar to that of Theorem 1.5.) Let C be a compact subset of S and (x_α) a net converging to 1. There exists a subnet (y_β) of (x_α) such that

$$\lim_{\beta} \mu(y_\beta C) = \underline{\lim}_{\alpha} \mu(x_\alpha C). \tag{1}$$

As in the proof of Theorem 1.5, let $h \in M(S)^*$ be chosen such that, for all real-valued $\nu \in M(S)$ we have that

$$-\ \|\nu\| \leq \underline{\lim}_{\beta}\ \nu(y_\beta C) \leq h(\nu) \leq \overline{\lim}_{\beta}\ \mu(y_\alpha C) \leq \|\nu\| \quad . \tag{2}$$

In particular

$$h(\mu) = \lim_{\beta} \mu(y_\beta C) . \tag{3}$$

Let $\varepsilon > 0$ be given. Then item (ii) implies that

$$U := \{u \in S : |h(\bar{u}*\mu) - h(\mu)| < \varepsilon\} \cap \{u \in S : |\bar{u}*\mu(C) - \mu(C)| < \varepsilon\}$$

is a neighbourhood of 1.

We can choose $u \in \mathrm{int}(U) \cap S_1$ and note that uU^{-1} is neighbourhood of 1. So eventually $(y_\beta) \subseteq uU^{-1}$. There exists a net$(u_\beta) \subseteq U$ such that eventually

$$y_\beta u_\beta = u. \tag{4}$$

Now from our definition of U, (4), (2), (3) and (1) (as in proof of Theorem 1.5) we have

$$\begin{aligned}\mu(C) &\leq \underline{\lim}_{\beta}\ \bar{u}_\beta*\mu(C) + \varepsilon \leq \underline{\lim}_{\beta}\ \bar{y}_\beta*(\bar{u}_\beta*\mu)(y_\beta C) + \varepsilon = \underline{\lim}_{\beta}\ \bar{u}*\mu(y_\beta C) + \varepsilon \\ &\leq h(\bar{u}*\mu) + \varepsilon \leq h(\mu) + 2\varepsilon = \lim_{\beta} \mu(y_\beta C) + 2\varepsilon = \underline{\lim}_{\alpha}\ \mu(x_\alpha C) + 2\varepsilon \\ &\leq \overline{\lim}_{\alpha}\ \mu(x_\alpha C) + 2\varepsilon \leq \mu(C) + 2\varepsilon.\end{aligned}$$

Hence $\lim_{\alpha} \mu(x_\alpha C) = \mu(C)$ and (i) is proved.

5.2 EXAMPLE. In general it is not true that $x \to \mu(xC)$ is a continuous map for all $\mu \in M_a(S)$ and compact $C \subseteq S$. For example take $S := [0,1]$ with the usual topology and multiplication given by

$$xy := \min(x+y,1) \quad \text{for all} \quad x,y \in S.$$

Then S is clearly a stip and $\bar{1} \in M_a(S)$. Let $C := \{0\}$, $x_n := 1 - \frac{1}{n}$ and note that $0 = \bar{1}(x_n C) \neq \bar{1}(1C) = 1$.

The results of this section are taken from G.L.G. Sleijpen ([92] and [93]).

6. SOME FURTHER NOTES

6.1. If the stip S is in fact a foundation semigroup then we recall from Theorem 2.2.6 that $M_a^\ell(S) = M_a^r(S)$. In that case results of this chapter remain true with $M_a(S)$ in place of $M_a^\ell(S)$. In fact this is also true for stips in general. We avoided giving this general form here for two reasons: Firstly the proof as known in the general case is rather complicated (see [94]), and secondly, as remarked in Chapter 3, a stip may after all be a foundation semigroup.

THEOREM (Sleijpen [94]). *For a stip* S *and* $\mu \in M(S)$, *we have the following items equivalent*:

(i) $\mu \in M_a(S)$;

(ii) ℓ_μ *is weakly continuous at* 1;

(iii) r_μ *is weakly continuous at* 1;

(iv) *both* ℓ_μ *and* r_μ *are norm continuous on* S.

6.2. We noted in Section 3 of Chapter 2 that the space of equi-regular measures is far from coinciding with the space of absolutely continuous measures for a very large class of locally compact topological semigroups. With this in mind, the next Theorem shows that a stip behaves just as well as one would expect, following the results of the preceding sections.

THEOREM (Dzinotyiweyi and Sleijpen [39]). *For a stip* S *we have that* $M_a(S) = M_e(S)$.

6.3. The reader who is interested in investigating whether every stip is a foundation semigroup is referred to the following partial result.

THEOREM (Sleijpen ([92] and [95])). *Let* S *be a stip such that* S_1 *is a* G_δ*-subset or* M(S) *is semisimple. Then* S *is a foundation semigroup.*

6 The convolution algebra $\ell^1(S)$

In this Chapter we prove certain results in which the algebra $\ell^1(S)$, first introduced by Hewitt and Zuckerman [60], plays a pivotal role. We do not give a systematic study of $\ell^1(S)$ as such. Most of the results we discuss have a significant algebraic flavour.

First we formally define the algebra $\ell^1(S)$.

Throughout this Chapter let S *be a discrete topological semigroup.*

Let $\ell^1(S)$ denote the set of all functions f in m(S) such that

$$\sum_{x \in S} |f(x)| < \infty .$$

With pointwise addition and scalar multiplication, with convolution

$$f*g(x) := \sum_{\substack{u,v \in S \\ uv = x}} f(u)g(v) \qquad (x \in S)$$

as product and with the norm

$$\|f\| = \|f\|_{\ell^1(S)} := \sum_{x \in S} |f(x)|,$$

$\ell^1(S)$ is a Banach algebra called the *discrete semigroup algebra* of S. One may identify $\ell^1(S)$ with the measure algebra M(S) and we shall exploit this identity when convenient.

1. REGULARITY OF MULTIPLICATION IN WEIGHTED SEMIGROUP ALGEBRAS

As remarked in Section 5 of Chapter 3, the property of regularity of multiplication in Banach algebras is invariant under passage both to subalgebras and to quotient algebras. In this section we study the regularity of multiplication in algebras related to the discrete semigroup algebra.

1.1 DEFINITION. By a *weight function* on S we mean a positive function $\omega : S \to \mathbb{R}$ such that

$$\omega(xy) \leq \omega(x)\omega(y) \quad (x,y \in S).$$

The set $\ell^1(S,\omega)$ of all complex-valued functions f on S such that $f\omega \in \ell^1(S)$, is a Banach algebra with convolution as product and norm given by

$$\|f\| := \|f\omega\|_{\ell^1(S)} .$$

If ω is a measurable weight function on a locally compact topological group G, then the set $L^1(G,\omega)$ of (equivalence classes of) all measurable complex-valued functions f on G such that $f\omega \in L^1(G)$, is a Banach algebra with convolution

$$f*g(x) := \int_G f(t)\, g(t^{-1}x)dm(t) \quad (x \in G)$$

as product, (where m is a given left Haar measure on G), and with the norm

$$\|f\| := \|f\omega\|_{L^1(G)} .$$

Corresponding to a weight function ω on S we define the function Ω on $S \times S$ by

$$\Omega(x,y) := \frac{\omega(xy)}{\omega(x)\ \omega(y)} .$$

We are now in a position to establish a criterion for the regularity of multiplication in the semigroup algebra $\ell^1(S,\omega)$.

1.2 THEOREM. *Let ω be a weight function on* S. *Then* (i) *implies* (ii), *and if* S *is cancellative* (i) *and* (ii) *are equivalent, where*

(i) $\ell^1(S,\omega)$ *has irregular multiplication*

(ii) *there exist sequences* $\{x_n\}$, $\{y_m\}$ *of distinct elements of* S *such that the double sequence* $\{\Omega(x_n,y_m)\}$ *has a non-zero repeated limit.*

Proof. First, one can easily note that $f \in \ell^1(S)$ if and only if $f\omega^{-1} \in \ell^1(S,\omega)$, and $h \in \ell^\infty(S)$ if and only if $h\omega \in \ell^\infty(S,\omega)$.

Now suppose item (i) holds. By the argument preceding Lemma 4.5.1, we can find sequences $\{f_n\}$ and $\{g_m\}$ in $\ell^1(S)$ and $h \in \ell^\infty(S)$ such that

$$a := \lim_n \lim_m h\omega(f_n\omega^{-1} * g_m\omega^{-1}) \text{ and } b := \lim_m \lim_n h\omega(f_n\omega^{-1} * g_m\omega^{-1})$$

exist and $a \neq b$. Now for any $f,g \in \ell^1(S)$ we have

$$h\omega(f\omega^{-1} * g\omega^{-1}) = \sum_{x,y \in S} f(x)g(y)\Omega(x,y)h(xy)$$

and so if we define $F : S \times S \to \mathbb{C}$ by

$$F(x,y) := \Omega(x,y)h(xy)$$

we have, in the notation of Appendix B.8, the map $F' : \ell^1(S) \times \ell^1(S) \to \mathbb{C}$ given by

$$F'(f,g) := \sum_{x,y \in S} f(x)g(y)\Omega(x,y)h(xy) = h\omega(f\omega^{-1} * g\omega^{-1}).$$

Hence by Appendices B.10 and B.7, we can find sequences $\{x_n\}$ and $\{y_m\}$ in S such that

$$a_1 := \lim_n \lim_m F(x_n,y_m) \quad \text{and} \quad b_1 := \lim_m \lim_n F(x_n,y_m)$$

exist and $a_1 \neq b_1$. So (passing to subsequences if necessary) we may assume that the x_n's and y_m's are distinct, and that both repeated limits of $\Omega(x_n,y_m)$ and of $h(x_n,y_m)$ exist. At least one repeated limit of $\Omega(x_n,y_m)$ must be non-zero (since $a_1 \neq b_1$), and item (ii) follows.

Next we assume that S is cancellative and (ii) holds. We may assume that

$$\lim_n \lim_m \Omega(x_n,y_m) > C > 0.$$

By taking a subsequence of $\{x_n\}$ if necessary, we further assume that

$$\lim_m (x_n,y_m) > C \quad (n \in \mathbb{N}).$$

Now arguing in a manner related but simpler than the proof of Lemma 4.3.1 we can find subsequences $\{u_n\}$ of $\{x_n\}$ and $\{v_m\}$ of $\{y_m\}$ such that

$$\{u_n v_m : n \leq m\} \cap \{u_n v_m : n > m\} = \phi$$

$$\text{and } \Omega(u_n,v_m) > C.$$

Let h be the characteristic function of the set $\{u_n v_m : n \leq m\}$. Then $h\omega \in \ell^\infty(S,\omega)$ and $\{\bar{u}_n \omega^{-1}\}$, $\{\bar{v}_m \omega^{-1}\}$ are bounded sequences in $\ell^1(S,\omega)$ such that

$$h\omega(\bar{u}_n \omega^{-1} * \bar{v}_m \omega^{-1}) = \begin{cases} \Omega(u_n, v_m) > C & \text{if } n \leq m \\ 0 & \text{if } n > m. \end{cases}$$

By the remarks preceding Lemma 4.5.1, it follows that multiplication in $\ell^1(S,\omega)$ is irregular. This completes our proof.

Next we mention some examples of weighted discrete semigroup algebras with regular multiplication.

1.3 COROLLARY.

(i) *For* $\alpha \geq 0$ *the function* ω_α *defined on the additive group* $\mathbf{Z}$ *by*

$$\omega_\alpha(n) := (1 + |n|)^\alpha \quad (n \in \mathbb{Z})$$

is a weight function. $\ell^1(\mathbb{Z},\omega_\alpha)$ *has regular multiplication if and only if* $\alpha > 0$.

(ii) *Let* ω *be a weight function on the additive semigroup of non-negative integers* $\mathbf{Z}^+$, *such that*

$$\lim_{n\to\infty} \frac{\omega(n+1)}{\omega(n)} = 0.$$

Then $\ell^1(\mathbf{Z}^+,\omega)$ *has regular multiplication.*

Proof. (i) It is trivial to note that ω_α is a weight function and that $\ell^1(\mathbb{Z},\omega_o) = \ell^1(\mathbb{Z})$ has irregular multiplication, by Theorem 4.5.4. Now for any $n \in \mathbb{N}$ and $\alpha > 0$ we have

$$\lim_{m\to\infty} \Omega_\alpha(n,m) = |n|^{-\alpha}(1 + |n|)^{-\alpha},$$

so that each repeated limit of $\{\Omega_\alpha(n,m)\}_{n,m}$ is zero. By Theorem 1.2 our result follows.

(ii) For any fixed $n \in \mathbb{N}$ we have that

$$\lim_{m\to\infty} \Omega(n,m) = \frac{1}{\omega(n)} \lim_{m\to\infty} \left[\frac{\omega(n+m)}{\omega(n+m-1)} \cdot \frac{\omega(n+m-1)}{\omega(n+m-2)} \cdots \frac{\omega(m+1)}{\omega(m)}\right] = 0.$$

Hence every repeated limit of Ω is zero and our result follows, by Theorem 1.2.

1.4 REMARK. Indeed one can construct weight functions on Z^+ which have the property mentioned in Corollary 1.3(ii). For instance ω given by

$$\omega(n) := e^{-n^2} \qquad (n \in Z^+)$$

is such a weight function.

The examples discussed above refer to countable semigroups. For such a class of semigroups we have the following general result.

1.5 COROLLARY. *Every countable semigroup* S *admits a weight function* ω *such that* $\ell^1(S,\omega)$ *has regular multiplication.*

Proof. Let F be the semigroup freely generated by a countable set $\{a_k : k \in \mathbb{N}\}$. Recalling that every $x \in F$ can be written uniquely as $x := a_{k_1} \ldots a_{k_r}$ we define $\omega : F \to \mathbb{R}^+$ by setting

$$\omega(x) := 1 + k_1 + \ldots + k_r.$$

Then ω is a weight function on F such that

$$\omega(xy) = \omega(x) + \omega(y) - 1 \qquad (x,y \in F).$$

Clearly $\omega(x_n) \to \infty$ for any sequence $\{x_n\}$ of distinct elements of F. So if $\{x_n\}$ and $\{y_m\}$ are sequences of distinct elements in F, we have that

$$\lim_{n\to\infty}\lim_{m\to\infty} \Omega(x_n,y_m) = \lim_{n\to\infty}\lim_{m\to\infty} \frac{\omega(x_n) + \omega(y_m) - 1}{\omega(x_n)\omega(y_m)} = 0$$

and similarly

$$\lim_{m\to\infty}\lim_{n\to\infty} \Omega(x_n,y_m) = 0.$$

By Theorem 1.2, it follows that $\ell^1(F,\omega)$ has regular multiplication.

Since S is countable, there is an epimorphism $\alpha : F \to S$. We now define $\omega' : S \to \mathbb{R}$ by

$$\omega'(x) := \min\{\omega(g) : g \in F \text{ and } \alpha(g) = x\}.$$

Then clearly ω' is a positive function.

For any $x,y \in S$ we can find $g_1, g_2 \in F$ such that

$$\alpha(g_1) = x, \quad \alpha(g_2) = y \text{ and so}$$

$$xy = \alpha(g_1)\alpha(g_2) = \alpha(g_1 g_2).$$

So

$$\omega'(xy) \leq \omega(g_1 g_2) \leq \omega(g_1)\omega(g_2)$$

and taking the minimum over all such g_1 and g_2 we get

$$\omega'(xy) \leq \omega'(x)\, \omega'(y).$$

Thus ω' is a weight function on S.

If $\{x_n'\}$ and $\{y_m'\}$ are sequences of distinct elements in S, we can find corresponding sequences of distinct elements $\{x_n\}$ and $\{y_m\}$ in F such that $\omega'(x_n') = \omega(x_n)$ and $\omega'(y_m') = \omega(y_m)$. So defining Ω' with respect to ω' as done for Ω with respect to ω, we have that

$$\Omega'(x_n', y_m') = \Omega(x_n, y_m) \to 0 \text{ as } n,m \to \infty.$$

Consequently $\ell^1(S,\omega)$ has regular multiplication, by Theorem 1.2.

1.6 THEOREM. *If G is a locally compact topological group, the following items are equivalent:*

(i) *G admits a weight function ω such that $L^1(G,\omega)$ has regular multiplication;*

(ii) *G is discrete and countable.*

Proof. That (ii) implies (i) is contained in Corollary 1.5. We now assume that (i) holds and prove (ii). Note that for any $x,y \in G$ we have

$$\omega(y^{-1})^{-1}\omega(y)^{-1} = \frac{\omega(xy)}{\omega(xy)\omega(y^{-1})\omega(y)} \leq \frac{\omega(xy)}{\omega(xyy^{-1})\omega(y)} = \Omega(x,y) \qquad (1)$$

and $G = \bigcup_{k=1}^{\infty} V_k$, where $V_k := \{y \in G : \frac{1}{k} \leq \omega(y^{-1})^{-1}\, \omega(y)^{-1}\}$.

Now suppose we have the case where G is discrete and take, on the contrary, G to be uncountable. Then, for some fixed $k \in \mathbb{N}$, we have V_k uncountable. Thus we can find a sequence $\{y_m\}$ of distinct elements of V_k. So if $\{x_n\}$ is any sequence in G, we have that

$$\frac{1}{k} \leq \Omega(x_n, y_m) \quad \text{for any } n,m \in \mathbb{N}; \quad \text{by (1)}.$$

Recalling Theorem 1.2, we have $\ell^1(G,\omega)$ irregular - contrary to assertion (i). By this contradiction we must have G countable.

Next, suppose we have the case where G is non-discrete. Since $G = \bigcup_{k=1}^{\infty} V_k$, there is a $k \in \mathbb{N}$ such that V_k has positive Haar measure. Let this k be fixed and choose $q \in d(V_k)$, and note that

$$1 \in d(V_k q^{-1}).$$

By Lemma 4.5.2, we can choose sequences $\{C_n\}$ and $\{D_m\}$ of compact sets with positive Haar measure such that

$$C_n, D_m \subset V_k q^{-1} \quad \text{and}$$

$$C_n D_m \cap C_i D_j = \phi \text{ for } n < m \text{ and } i > j.$$

So

$$D_m q \subset V_k \quad \text{and}$$

$$C_n(D_m q) \cap C_i(D_j q) = (C_n D_m \cap C_i D_j)q = \phi \text{ for } n < m \text{ and } i > j.$$

Taking μ to be Haar measure on G, let

$$f_n := \mu(C_n)^{-1} \chi_{C_n}, \quad g_m := \mu(D_m q)^{-1} \chi_{D_m q} \quad \text{and} \quad h := \chi_A$$

$$\text{where } A := \bigcup_{m=1}^{\infty} \bigcup_{n>m} C_n D_m q.$$

Then $\{f_n \omega^{-1}\}$ and $\{g_m \omega^{-1}\}$ are bounded sequences in $L^1(G,\omega)$ and $h\omega \in L^1(G,\omega)^*$.

From the definition of V_k and item (1), we have that

$$\Omega(x,y) \geq \frac{1}{k} \quad \text{for all } x \in G, \; y \in D_m q \text{ and } m \in \mathbb{N}.$$

Consequently

$$h\omega(f_n\omega^{-1} * g_m\omega^{-1}) = \int_{C_n}\int_{D_m q} \Omega(x,y)h(xy)\,d\mu(x)\,d\mu(y) = \begin{cases} a \geq \frac{1}{k} & \text{if } n > m \\ 0 & \text{if } n < m. \end{cases}$$

Recalling the remarks preceding Lemma 4.5.1, we have that $L^1(G,\omega)$ has irregular multiplication. By this conflict we cannot have G non-discrete. This completes our proof.

1.6 Notes. The results of this section are taken from the paper of Craw and Young [23]. There does not seem to be much work done on weighted group or semigroup algebras.

Recently W.G. Bade and H.G. Dales [3] have proved some results on weighted semigroup algebras defined on the underlying semigroup $[0,\infty)$. We feel it would be a useful contribution to extend some of their results to the weighted algebras $\ell^1(S,\omega)$ and $L^1(G,\omega)$.

2. EXISTENCE OF APPROXIMATE IDENTITIES

In this section we establish criteria for the existence of an approximate identity of the discrete semigroup algebra of an inverse semigroup.

2.1 DEFINITION. An *inverse semigroup* is a semigroup S such that for each $s \in S$ there exists a unique $s^{\bullet} \in S$ with

$$ss^{\bullet}s = s \quad \text{and} \quad s^{\bullet}ss^{\bullet} = s^{\bullet} .$$

2.2 EXERCISE. If S is an inverse semigroup then E_S is a commutative subsemigroup.

We refer the reader to the book of Clifford and Preston [17] for more details on inverse semigroups.

Since elements of $\ell^1(S)$ are arbitrarily small outside finite sets, Proposition 2.2.9 can be formulated for $\ell^1(S)$ as follows.

2.3 PROPOSITION. *The algebra* $\ell^1(S)$ *admits a bounded right approximate identity with bound* M *if and only if given* $\varepsilon > 0$ *and any finite number of elements* $s_1, s_2, \ldots, s_n$ *in* S, *there exists* $\alpha \in \ell^1(S)$ *such that*

$$\|\alpha\| \leq M \quad \textit{and} \quad \|\bar{s}_i - \alpha * \bar{s}_i\| < \varepsilon \quad \textit{for} \quad i = 1,2,\ldots,n.$$

Towards our main result, (Theorem 2.7), we prove some lemmas.

2.4 LEMMA. *Let* S *be an inverse semigroup and* $e_1,\ldots,e_k \in E_S$ *be given. Then there exists* $\alpha_k \in \ell^1(E_S)$ *such that*

$$\|\alpha_k\| \leq 2^{k-1} - 1 \quad \textit{and} \quad \alpha_k * \bar{e}_j = \bar{e}_j \quad \textit{for} \quad j = 1,\ldots,k.$$

Proof. Recalling that E_S is a commutative semigroup an easy induction argument on k with

$$\alpha_1 := \bar{e}_1 \quad \text{and} \quad \alpha_k := \alpha_{k-1} + \bar{e}_k - \bar{e}_k * \alpha_{k-1}$$

yields the result.

2.5 DEFINITION. Let S be an inverse semigroup and k a positive integer. We say E_S satisfies condition (D_k) if given $u_1,\ldots,u_{k+1} \in E_S$ we can find $e \in E_S$ and integers i,j such that

$$1 \leq i < j \leq k+1, \quad eu_i = u_i \quad \text{and} \quad eu_j = u_j.$$

2.6 LEMMA. *If* S *is an inverse semigroup, then* $\ell^1(E_S)$ *admits a bounded approximate identity if and only if* E_S *satisfies condition* (D_k) *for some positive integer* k.

Proof. Suppose E_S satifies condition (D_k) for some positive integer k and let $\{u_1,\ldots,u_n\}$ be any finite subset of E_S. Then condition (D_k) implies the existence of $e_1,\ldots,e_k \in E_S$ such that for each $i \in \{1,\ldots,n\}$ there exists $j \in \{1,\ldots,k\}$ with $e_j u_i = u_i$. Then choosing $\alpha_k \in \ell^1(E_S)$ as in Lemma 2.4 we get

$$\bar{u}_i = \overline{e_j u_i} = \bar{e}_j * \bar{u}_i = \alpha_k * \bar{e}_j * \bar{u}_i = \alpha_k * \overline{e_j u_i} = \alpha_k * \bar{u}_i \quad \text{for } i = 1,\ldots,n.$$

Now recalling Proposition 2.3 (and the commutativity of E_S) we conclude that $\ell^1(E_S)$ has a bounded approximate identity with bound 2^k-1.

Conversely suppose that $\ell^1(E_S)$ admits a bounded approximate identity with bound M (say). Let $e_1,\ldots,e_{k+1} \in E_S$ be given, where k is an integer such that $k \geq M$. Proposition 2.3 says that we can find $\alpha \in \ell^1(E_S)$ such that

$$\|\alpha\| \leq M \quad \text{and} \quad \|\bar{e}_i - \alpha * \bar{e}_i\| < \frac{1}{k+1} \quad \text{for} \quad i = 1,\ldots,k+1.$$

With obvious notation we write

$$\alpha = \sum_r \lambda_r \bar{u}_r$$

where $u_r \in E_S$ and $\lambda_r \in \mathbb{C}$, for all r. For each $i \in \{1,\ldots,k+1\}$ let $A_i := \{r : e_i u_r = e_i\}$ and hence note that

$$1 - \sum_{r\in A_i} |\lambda_r| \leq |1 - \sum_{r\in A_i} \lambda_r| = \|\bar{e}_i - \sum_{r\in A_i} \lambda_r \bar{u}_r * \bar{e}_i\|$$

$$\leq \|\bar{e}_i - \alpha * \bar{e}_i\| < \frac{1}{k+1} .$$

So

$$\sum_{r\in A_i} |\lambda_r| > 1 - \frac{1}{k+1} = \frac{k}{k+1} .$$

If the sets $A_1,\ldots,A_{k+1}$ were pairwise disjoint we would get

$$M \geq \sum_r |\lambda_r| \geq \sum_{i=1}^{k+1} \sum_{r\in A_i} |\lambda_r| > (k+1) \frac{k}{k+1} = k$$

which contradicts our choice of k. Hence $A_i \cap A_j \neq \phi$ for some i,j with $1 \leq i < j \leq k + 1$. Thus E_S satisfies condition (D_k).

2.7 THEOREM. *Let* S *be an inverse semigroup. Then the following items are equivalent:*

(i) $\ell^1(S)$ *admits a bounded approximate identity;*

(ii) $\ell^1(S)$ *admits a bounded left approximate identity;*

(iii) $\ell^1(E_S)$ *admits a bounded left approximate identity;*

(iv) E_S *satisfies condition* (D_k) *for some positive integer* k.

Proof. That (i) implies (ii) is trivial. Recalling that E_S (and hence $\ell^1(E_S)$) is commutative, the equivalence of (iii) and (iv) is established in Lemma 2.6.

We now prove the equivalence of (ii) and (iii).

Firstly, suppose $\ell^1(S)$ admits a bounded left approximate identity with bound M. Let $T : S \to E_S$ be given by $T(s) := s^{\cdot}s$. Then, since E_S is commutative we have, for all $s \in S$ and $e \in E_S$,

$$T(se) := (se)^{\cdot}(se) = e^{\cdot}s^{\cdot}se = e(s^{\cdot}s)e = (s^{\cdot}s)e = T(s)e.$$

Hence T extends to a norm-decreasing linear map (which we also denote by) $T : \ell^1(S) \to \ell^1(E_S)$ such that

$$T(\alpha * \bar{e}) = T(\alpha) * \bar{e} \quad (\alpha \in \ell^1(S) \text{ and } e \in E_S).$$

Let $e_1, \ldots, e_n \in E_S$ and $\varepsilon > 0$ be given, there exists $\alpha \in \ell^1(S)$ such that

$$\|\alpha\| \leq M \text{ and } \|\bar{e}_i - \alpha * \bar{e}_i\| < \varepsilon \quad \text{for } i = 1, \ldots, n.$$

Then $T(\alpha) \in \ell^1(E_S)$, $\|T(\alpha)\| \leq M$ and

$$\|\bar{e}_i - T(\alpha) * \bar{e}_i\| = \|T(\bar{e}_i) - T(\alpha * \bar{e}_i)\| \leq \|\bar{e}_i - \alpha * \bar{e}_i\| < \varepsilon .$$

Hence $\ell^1(E_S)$ has a bounded left approximate identity.

Secondly, suppose $\ell^1(E_S)$ admits a bounded left approximate identity with bound M (say). Let $s_1, \ldots, s_n \in S$ and $\varepsilon > 0$ be given. Since $s_i s_i^{\cdot} \in E_S$, there exists $\alpha \in \ell^1(E_S)$ such that

$$\|\alpha\| \leq M \text{ and } \|\overline{s_i s_i^{\cdot}} - \alpha * \overline{s_i s_i^{\cdot}}\| < \varepsilon \quad \text{for } i = 1, \ldots, n.$$

Thus for $i = 1, \ldots, n$; as $s_i = s_i s_i^{\cdot} s_i$, we have that

$$\|\bar{s}_i - \alpha * \bar{s}_i\| = \|(\overline{s_i s_i^{\cdot}} - \alpha * \overline{s_i s_i^{\cdot}}) * \bar{s}_i\|$$

$$\leq \|\overline{s_i s_i^{\cdot}} - \alpha * \overline{s_i s_i^{\cdot}}\| .$$

Hence $\ell^1(S)$ admits a bounded left approximate identity.

Thus (ii) and (iii) are equivalent. Since $\ell^1(E_S)$ is commutative we have (also shown) that (i) and (iii) are equivalent.

From the definition of a Brandt semigroup S, (see item 4.4(b)(i)), we note that $ef \notin \{e,f\}$ for distinct non-zero elements $e,f \in E_S$. So, if E_S is infinite, E_S does not have property (D_k) for positive integers k. Hence Theorem 2.7 implies

2.8 COROLLARY. *If* S *is a Brandt semigroup with* E_S *infinite, then* $\ell^1(S)$ *does not have a bounded left approximate identity.*

The proof of the following result is contained in that of Theorem 2.7.

2.9 PROPOSITION. *If* S *is an inverse semigroup, then* $\ell^1(S)$ *admits a right approximate identity of bound* 1 *if and only if* E_S *is directed upwards.*

The results of this section are taken from the paper of Duncan and Namioka [28]. Theorem 2.7 and Proposition 2.9 are also related to the results of Charles Lahr [69].

3. A FACTORIZABLE BANACH ALGEBRA WITHOUT AN APPROXIMATE IDENTITY

We recall that a Banach algebra A is said to be factorizable if every element $a \in A$ can be written as $a = bc$ for some $b,c \in A$. It is well known that a Banach algebra with a bounded approximate identity is factorizable - recall Theorem 2.2.11. For some time it was not clear whether, conversely, factorization implies the existence of a bounded approximate identity. Paschke [78] showed that this converse does not hold. In this section, following Leinert [72], we construct a commutative semigroup S such that $\ell^1(S)$ is factorizable but does not have an (even unbounded) approximate identity.

Let S be the semigroup (with pointwise addition) of all real sequences $x := \{x_n\}$ with $x_n > 0$ for all but finitely many $n \in \mathbb{N}$. Consider any $f \in \ell^1(S)$. We can find sequences $a_n := \{a_{n,k}\}_{k\in\mathbb{N}}$ and $\{\lambda_n\}$ such that $\lambda_n \in \mathbb{C}$, $a_n \in S$ $(n \in \mathbb{N})$ and

$$f = \sum_{n=1}^{\infty} \lambda_n \chi_{\{a_n\}}.$$

For each $k \in \mathbb{N}$, let $A_k := \{a_{n,k} : 1 \leq n \leq k \text{ and } a_{n,k} > 0\}$ and

$$h_k := \begin{cases} \frac{1}{2}\min(A_k) & \text{if } A_k \neq \phi. \\ 1 & \text{if } A_k = \phi. \end{cases}$$

Then $h := \{h_k\} \in S$.

Now consider the sequence $g_n := a_n - h$. The kth entry of g_n is

$$a_{n,k} - h_k \begin{cases} > 0 & \text{if} \quad A_k \neq \phi, \\ \leq 0 & \text{if} \quad A_k = \phi. \end{cases}$$

By definition of A_k, we have $A_k \neq \phi$ for all but a finite number of $k \in \mathbb{N}$. Thus $g_n \in S$, for all $n \in \mathbb{N}$.

Next we note that, that $g_n = a_n - h$ implies $f = \chi_{\{h\}} * \sum_{n=1}^{\infty} \lambda_n \chi_{\{g_n\}}$; so f factorizes. Hence $\ell^1(S)$ is factorizable.

For every $g \in S$ and $f \in \ell^1(S)$ we have g not contained in the support of $\chi_{\{g\}} * f$ (by our definition of elements of S) and so

$$\|\chi_{\{g\}} * f - \chi_{\{g\}}\| \geq 1.$$

Hence $\ell^1(S)$ has no approximate unit.

The reader may also note that $\ell^1(S)$ is a subalgebra of a commutative group algebra and so $\ell^1(S)$ is a semisimple algebra.

4. AMENABILITY OF THE ALGEBRA $\ell^1(S)$

Let S be any discrete topological semigroup. For $s \in S$, $f \in \ell^\infty(S \times S)$ and $M \in \ell^\infty(S \times S)^*$ we define $s.f$, $f.s \in \ell^\infty(S \times S)$ and $s.M$, $M.s \in \ell^\infty(S \times S)^*$ by

$$s.f(t,t') := f(st,t'), \quad f.s(t,t') := f(t,t's)$$

$$s.M(f) := M(s.f) \quad \text{and} \quad M.s(f) := M(f.s).$$

Let $\pi : \ell^1(S \times S) \to \ell^1(S)$ be given by $\pi(\bar{s},\bar{t}) := \overline{st}$ for all $s,t \in S$. For $\phi \in \ell^\infty(S)$ we define $\pi^*\phi \in \ell^\infty(S \times S)$ and $\pi^{**}M \in \ell^\infty(S)^*$ by

$$\pi^*\phi(\tau) := \phi(\pi(\tau)) \quad \text{and} \quad \pi^{**}M(\phi) := M(\pi^*\phi)$$

for all $\tau \in \ell^1(S \times S)$.

An element M of $\ell^\infty(S \times S)^*$ is called a *virtual diagonal* for $\ell^1(S)$ if, for all $s \in S$, we have

$$s.M = M.s \quad \text{and} \quad (\pi^{**}M)_s = \bar{s}.$$

$\ell^1(S)$ is said to be *amenable* if there exists a virtual diagonal for $\ell^1(S)$.

The semigroup S is said to be *amenable* if there exists a left and right invariant mean on $m(S) = \ell^{\infty}(S)$. (Note that existence of a left invariant mean and a right invariant on $\ell^{\infty}(S)$ implies the existence of a left and right invariant mean on $\ell^{\infty}(S)$ - see e.g. Day [25].)

Let θ denote the usual canonical projection of $\ell^1(S \times S)$ into $\ell^{\infty}(S \times S)^*$. An *approximate diagonal* for $\ell^1(S)$ is a bounded net (m_α) in $\ell^1(S \times S)$ such that, for all $s \in S$,

$$\lim_\alpha (s.\theta(m_\alpha) - \theta(m_\alpha).s) = 0 \quad \text{and} \quad \lim_\alpha \pi(m_\alpha)*\bar{s} = s.$$

We then have

4.1 LEMMA. *The algebra $\ell^1(S)$ has an approximate diagonal if and only if it has a virtual diagonal.*

Proof. Suppose $\ell^1(S)$ has an approximate diagonal (denoted by) (m_α). Let M be a weak*-cluster point of $\theta(m_\alpha)$ in $\ell^{\infty}(S \times S)^*$. We may assume that $M = \text{weak*-}\lim_\alpha \theta(m_\alpha)$. Then, for all f in $\ell^{\infty}(S \times S)$, we have that

$$\begin{aligned} s.M(f) := M(s.f) &= \lim_\alpha \theta(m_\alpha)\ (s.f) \\ &= \lim_\alpha s.\theta(m_\alpha)(f) \\ &= \lim_\alpha \theta(m_\alpha).s(f) \\ &= \lim_\alpha \theta(m_\alpha)(f.s) \\ &= M(f.s) = M.s(f) \end{aligned}$$

and

$$\begin{aligned} (\pi^{**}M)_s(\phi) := \pi^{**}M(\phi_s) &= M(\pi^*\phi_s) \\ &= \lim_\alpha \theta(m_\alpha)(\pi^*\phi_s) \\ &= \lim_\alpha \pi^*\phi_s(m_\alpha) \end{aligned}$$

$$= \lim_{\alpha} \phi(\pi(m_\alpha)*\bar{s})$$

$$= \phi(\bar{s}) = \theta(\bar{s})(\phi)$$

for all $\phi \in \ell^\infty(S)$. Hence $(\pi^{**}M)_s = \bar{s}$ (on identifying $\theta(\bar{s})$ with $\bar{s}$).

Thus M is a virtual diagonal for $\ell^1(S)$.

By arguing in a related fashion the reader should be able to handle the converse case.

4.2 PROPOSITION. *Let* S *be an inverse semigroup with an identity element* 1. *Then the following items are equivalent:*

(i) *There exists* M *in* $\ell^\infty(S \times S)^*$ *such that*

$$M(1) = 1 \quad \textit{and} \quad s.M = M.s \quad \textit{for all} \quad s \in S;$$

(ii) S *is amenable.*

Proof. Suppose (i) holds. We define m in $\ell^\infty(S)^*$ by

$$m(f) := M(\bar{f})$$

where $\bar{f}(s,t) := f(s)$, for all f in $\ell^\infty(S)$ and s,t in S. Then, for all x in S, we have that

$$\begin{aligned} m({}_x f) := M({}_x\bar{f}) = M(x.\bar{f}) = x.M(\bar{f}) &= M.x(\bar{f}) \\ &= M(\bar{f}.x) \\ &= M(\bar{f}) \\ &= m(f) . \end{aligned}$$

Also $m(1) = M(1) = 1$. Thus m is a left invariant mean on $\ell^\infty(S)$. Similarly, there exists a right invariant mean on $\ell^\infty(S)$. So S is amenable.

Conversely, suppose (ii) holds. Let m be an invariant mean on $\ell^\infty(S)$ and define $M \in \ell^\infty(S \times S)^*$ by

$$M(f) := \int_S f(x^\bullet, x)\,dm(x) \quad \text{for all } f \text{ in } \ell^\infty(S \times S).$$

Clearly $M(1) = m(1) = 1$. For each s in S and f in $\ell^\infty(S \times S)$ we have

$$\begin{aligned}
s.M(f) := M(s.f) :=& \int_S s.f(x^\bullet, x)\,dm(x) \\
=& \int_S f(sx^\bullet, x)\,dm(x) \\
=& \int_S f(s(xs)^\bullet, xs)\,dm(x) \\
=& \int_S f(ss^\bullet x^\bullet, xs)\,dm(x) \\
=& \int_S f((xss^\bullet)^\bullet, (xss^\bullet)s)\,dm(x) \\
=& \int_S f(x^\bullet, xs)\,dm(x) \\
=& \int_S f.s(x^\bullet, x)\,dm(x) \\
=& M(f.s) = M.s(f).
\end{aligned}$$

Thus $s.M = M.s$.

4.3 REMARKS.

(a) One can easily note (from the proof of Proposition 4.2) that S is amenable if $\ell^1(S)$ is amenable, for any discrete semigroup S.

(b) In the case of a discrete amenable group G with an invariant mean m, we have $\ell^1(G)$ amenable; in fact the virtual diagonal M can be explicitly defined by

$$M(f) := m(x \to f(x, x^{-1})) \text{ for all } f \text{ in } \ell^\infty(G \times G).$$

[See proof of Proposition 4.2.]

We will now discuss the amenability of $\ell^1(S)$ for a special class of inverse semigroups - namely Brandt semigroups. Our reference for the algebraic results on Brandt semigroups is the book of Clifford and Preston [20].

4.4 DEFINITIONS. (a) A *principal series* of a semigroup S is a chain

$$S_1 \supset S_2 \supset \dots \supset S_m \supset S_{m+1}$$

of ideals S_i of S, beginning with $S_1 := S$ and ending with $S_{m+1} := \phi$, such that there is no ideal of S between S_i and S_{i+1}; for $i = 1,\dots,m$.

(b) By a *Brandt semigroup* we mean a semigroup S with a zero element 0 and satisfying the following axioms:

(i) To each non-zero element x of S there corresponds unique elements e,f and x' in S such that

$$ex = x, \quad xf = x \quad \text{and} \quad x'x = f.$$

(ii) If e and f are non-zero idempotents of S, then $eSf \neq \{0\}$.

We recall that a Brandt semigroup S over a group G with index set I consists of elementary $I \times I$ - matrices over $G \cup \{0\}$ and a zero matrix θ - see e.g. [20, page 102]. Writing $S = \{(g)_{ij} : g \in G;\ i,j \in I\} \cup \{\theta\}$, where $(g)_{ij}$ is the matrix with (k,ℓ) - entry equal to g if $(k,\ell) = (i,j)$ and 0 if $(k,\ell) \neq (i,j)$, we get

$$(g)_{ij} \cdot (h)_{k\ell} = \begin{cases} (gh)_{i\ell} & \text{if } j = k, \\ \theta & \text{if } j \neq k. \end{cases}$$

With the help of item 4.1, the reader should be able to solve the following exercises. (See also Johnson [64] for more general results.)

4.5 EXERCISES

(i) Let T be an ideal of S and $\ell^1(S)$ be amenable. If H is a semigroup such that $\ell^1(H)$ and $\ell^1(S)/\ell^1(T)$ are isomorphic, we have the algebra $\ell^1(H)$ amenable.

(ii) If H and S are semigroups such that $\ell^1(H)$ and $\ell^1(S)$ are amenable, then $\ell^1(H \times S)$ is amenable.

(iii) Let J be a Brandt semigroup with finite index set I over a trivial group. Writing $J = \{e_{ij} : i,j \in I\} \cup \{\theta\}$, we have that

$$M := \sum_{i \in I} ((\bar{e}_{ii_o} \times \bar{e}_{i_o i}) - (\bar{\theta} \times \bar{e}_{i_o i}) - (\bar{e}_{ii_o} \times \bar{\theta})) + (n+1)(\bar{\theta} \times \bar{\theta})$$

is a virtual diagonal for $\ell^1(J)$, where $i_o \in I$ is fixed and $n := |I|$.

4.6 THEOREM. *Let* S *be a Brandt semigroup over a group* F *with a finite index set* I. *Then* $\ell^1(S)$ *is amenable if and only if* G *is amenable.*

Proof. We assume the notation on components of S introduced just after Devinition 4.4(b).

First, let G be amenable. Then $\ell^1(G)$ is amenable, as noted in Remark 4.3(b). Let J be the Brandt semigroup with index set I over the trivial group. Then, from Exercises 4.5(iii) and (ii), we have that $\ell^1(J)$ and hence $\ell^1(J \times G)$ is amenable. Noting that $T := \{(\theta, g) : g \in G\}$ is an ideal of $J \times G$ one can formulate a standard argument showing that $\ell^1(S)$ is isometrically isomorphic with $\ell^1(J \times G)/\ell^1(T)$. Hence, recalling Exercise 4.5(i), we have that $\ell^1(S)$ is amenable.

Next suppose that $\ell^1(S)$ is amenable with virtual diagonal M. Let $j \in J$ be fixed. Corresponding to each $f \in \ell^\infty(G \times G)$ we define $\tilde{f}$ on $S \times S$ by

$$\tilde{f}(s,t) := \begin{cases} f(g,h) & \text{if } (s,t) = ((g)_{ij}, (h)_{ij}) \\ 1 & \text{otherwise.} \end{cases}$$

For $u \in G$ we have

$$\begin{aligned} (u)_{ij} \cdot \tilde{f}(s,t) &= \tilde{f}((u)_{ij}\, s,t) \\ &= \begin{cases} \tilde{f}((ug)_{ij}, (h)_{ij}) & \text{if } (s,t) = ((g)_{jj}, (h)_{jj}) \\ 1 & \text{otherwise.} \end{cases} \\ &= \begin{cases} f(ug), h) & \text{if } (s,t) = ((g)_{jj}, (h)_{jj}) \\ 1 & \text{otherwise.} \end{cases} \\ &= \widetilde{u.f}(s,t). \end{aligned}$$

Thus $(u)_{ij}.\tilde{f} = \widetilde{u.f}$. Similarly $\tilde{f}.(u)_{jj} = \widetilde{f.u}$. Now let m be defined on $\ell^\infty(G \times G)$ by $m(f) := M(\tilde{f})$.

Then, for all $f \in \ell^\infty(G \times G)$ and $u \in G$, we have

$$u.m(f) := m(u.f) := M(\widetilde{u.f}) = M((u)_{jj}.\tilde{f})$$

$$= M(\tilde{f}.(u)_{jj}) \quad \text{since M is a virtaul diagonal}$$

$$= M(\widetilde{f.u})$$

$$= m(f.u).$$

Also $m(1) = M(1) = 1$. Recalling Proposition 4.2, we have G amenable.

4.7 THEOREM. *Let S be an inverse semigroup with E_S finite. Then $\ell^1(S)$ is amenable if and only if each maximal subgroup of S is amenable.*

Proof. Since E_S is finite, S has a principal series

$$S = S_1 \supset S_2 \supset \ldots \supset S_m \supset S_{m+1} = \phi$$

of ideals of S. We have S_m/S_{m+1} (i.e. S_m) a simple inverse semigroup with a finite number of idempotents and so is a completely simple inverse semigroup, i.e. a group. For $i = 1,\ldots,m-1$, we have S_i/S_{i+1} 0-simple with a finite number of idempotents and so is a completely 0-simple inverse semigroup, i.e. a Brandt semigroup [20]. Hence, recalling Exercise 4.5(i), we have that $\ell^1(S)$ is amenable if and only if $\ell^1(S_i/S_{i+1})$ is amenable, for $i = 1,2,\ldots,m$.

Let G_m be the group S_m/S_{m+1} and G_i the group of the Brandt semigroup S_i/S_{i+1} for $i = 1,\ldots,m-1$. Recalling Theorem 4.6, we thus have $\ell^1(S_i/S_{i+1})$ amenable if and only if G_i is amenable. Up to isomorphism the groups G_i are precisely the maximal subgroups of S. Hence our result follows.

Next we show that the conclusion of Theorem 4.6 collapses if the index set I is infinite.

4.8 THEOREM. *Let S be a Brandt semigroup with an infinite index set over an arbitrary group. Then $\ell^1(S)$ is not amenable.*

Proof. Suppose, on the contrary, $\ell^1(S)$ is amenable. Then $\ell^1(S)$ has a virtual diagonal, by Lemma 4.1. In particular $\ell^1(S)$ has a bounded left approximate identity. This contradicts Corollary 2.8. So our Theorem follows.

That a discrete group G is amenable if and only if $\ell^1(G)$ is amenable was proved by Johnson [64]. The results of this section are taken from the paper of Duncan and Namioka [28].

7 Non-Archimedean Fourier theory

In this chapter we depart from the frame of reference of the previous chapters by replacing the Archimedean fields $\mathbb{R}$ and $\mathbb{C}$ with a non-Archimedean field K. An excellent reference for such studies in Functional analysis, which we will follow closely, is the book of A.C.M. van Rooij [102] - in fact the reader may find it useful to study this chapter with the book [102] on the same desk. The results in this chapter are taken from the paper of Dzinotyiweyi and van Rooij [40]. As in the case with most results in non-Archimedean harmonic analysis, these are not results to stand on their own; accordingly, we begin with a summary of some standard results and observations.

1. PREREQUISITES

1.1 DEFINITIONS. Let K be a field. A *non-Archimedean valuation* on K is a function $|\ | : K \to \mathbb{R}$ with the following properties:

(a) $|x| \geq 0$

(b) $|x| = 0$ if and only if $x = 0$

(c) $|x+y| \leq \max\{|x|,|y|\}$

(d) $|xy| = |x||y|$

for all x and y in K. $(K,|\ |)$ is called a *non-Archimedean valued field*. Evidently the valuation $|\ |$ induces a metric on K given by

$$\rho(x,y) := |x-y| \quad \text{for all} \quad x,y \in K.$$

If (K,ρ) is a complete metric space we say $(K,|\ |)$ is a *complete non-Archimedean valued field*. For convenience we shall simply write K in place of $(K,|\ |)$. Also K is assumed to be complete with nontrivial valuation in this chapter.

R := $\{x \in K : |x| \leq 1\}$ is a subring of K and I := $\{x \in K : |x| < 1\}$ a maximal ideal of R. The quotient $k := R/I$ is called the *residue class field* of K and we denote the natural homomorphism $R \to k$ by $x \to \bar{x}$ $(x \in R)$.

The letter p will denote the characteristic of (the residue class field) k while the group of all roots of unity of K whose orders are not divisible by p is denoted by T_K. The elements x,y of T_K are such that $|x-y| = 1$ if $x \neq y$, so the natural topology on T_K is discrete.

We also denote the unit element of K by 1 and the K-valued characteristic function of a set X by χ_X.

1.2 BANACH SPACES. A *non-Archimedean norm* on a vector space E over K is a function $\| \ \| : E \to \mathbb{R}$ with the properties:

(a) $\|x\| \geq 0$

(b) $\|x\| = 0$ if and only if $x = 0$

(c) $\|x+y\| \leq \max\{ \|x\| , \|y\| \}$

(d) $\|\lambda x\| = |\lambda| \, \|x\|$

for all x and y in E and λ in K. If E is complete with respect to the metric induced by $\| \ \|$, then E is called a *non-Archimedean Banach space* (over K with the norm $\| \ \|$) and the dual of E, denoted by E*, is the space of all continuous linear maps $E \to K$.

If, in addition, E is an algebra the norm $\| \ \|$ is said to be *power multiplicative* if $\|x^n\| = \|x\|^n$ for all $x \in E$ and $n \in \mathbb{N}$. For commutative E, this is equivalent to : $\|x^2\| = \|x\|^2$ $(x \in E)$. (See e.g. [102, page 223].) One of the important results on power multiplicative Banach algebras now follows.

1.3 THEOREM (Springer [102, page 224]). *Let* A *be a commutative Banach algebra with power multiplicative norm. Then for every* $a \in A$ *there exists a Banach algebra homomorphism* ϕ *of* A *into some complete valued field* L *that contains* K, *such that* $|\phi(a)| = \|a\|$.

1.4 DEFINITION. A subset X of a non-Archimedean Banach space E is said to be *orthonormal* if for every finite sequence $\{x_1,\ldots,x_n\}$ in X and $\{\lambda_1,\ldots,\lambda_n\}$ in K we have

$$\|\lambda_1 x_1 + \ldots + \lambda_n x_n\| = \max\{|\lambda_1|,\ldots,|\lambda_n|\} \quad \text{and}$$

$$\|x\| = 1 \quad \text{for all} \quad x \text{ in } X.$$

1.5 SPACES OF FUNCTIONS. Let X be any set.

The bounded maps $f : X \to K$ form a linear space $\ell^\infty(X)$, which is a Banach space under the norm $\| \|_\infty$ given by

$$\|f\|_\infty := \sup\{|f(x)| : x \in X\}.$$

$c_o(X) := \{f \in \ell^\infty(X) :$ for every $\varepsilon > 0$, the set $\{x \in X : |f(x)| \geq \varepsilon\}$ is finite$\}$ is a closed linear subspace of $\ell^\infty(X)$.

Now suppose X is endowed with a topology. The bounded continuous functions $X \to K$ form a closed linear subspace, $C(X)$, of $\ell^\infty(X)$. For a locally compact X, let $C_o(X) := \{f \in C(X) :$ for every $\varepsilon > 0$, the set $\{x \in X : |f(x)| \geq \varepsilon\}$ is compact$\}$.

Recall that a subset E of X is said to be *clopen* if it is both closed and open in X. Now for any topological space X and distinct a,b in X suppose $f : X \to K$ is a continuous function such that $f(a) \neq f(b)$. Then setting $\varepsilon := \frac{1}{2}|f(a)-f(b)|$, we see that $\{x \in X : |f(x)-f(a)| \leq \varepsilon\}$ is a clopen subset of X that contains a but not b - see [102, pages 22 and 25]. Hence the continuous functions from X into K separate the points of X if and only if the clopen subsets of X separate points. This justifies the zero-dimensionality requirement in Blanket assumption 1.8.

The following result is the natural version of the Stone-Weierstrass theorem for the non-Archimedean case.

1.6 THEOREM (Kaplansky [102, page 218]). *Let X be a locally compact zero-dimensional topological space and A a closed subalgebra of* $C_o(X)$ *such that for any distinct points* x *and* y *in* X *we have* $f(x) = 0$ *and* $f(y) = 1$ *for some* f *in* A. *Then* $A = C_o(X)$.

1.7 TIGHT MEASURES. Let X be a zero-dimensional topological space and $\mathcal{B}(X)$ the ring of all clopen subsets of X. A *tight measure* on X is a map $\mu : \mathcal{B}(X) \to K$ with the properties:

(a) μ is additive. (Hence $\mu(\phi) = 0$.)

(b) For all A in $\mathcal{B}(X)$, the set $\{\mu(B) : B \subset \mathcal{B}(X)$ and $B \subset A\}$ is bounded.

(c) If $\mathcal{A} \subset \mathcal{B}(X)$ is shrinking (i.e. the intersection of any two sets in $\mathcal{A}$ contains a set in $\mathcal{A}$) and $\cap\mathcal{A} = \phi$, then $\lim_{A\in\mathcal{A}} \mu(A) = 0$.

For each tight measure μ, $A \in \mathcal{B}(X)$ and $x \in X$, let

$$|\mu|(A) := \sup\{|\mu(B)| : B\in\mathcal{B}(X),\ B \subseteq A\} \text{ and } N_\mu(x) := \inf\{|\mu|(O): O\in\mathcal{B}(X),\ x \in O\}$$

The support of μ, *supp*(μ), is the closure of the set $\{x \in X : N_\mu(x) > 0\}$. The set, $M(X)$, of all tight measures on X forms a non-Archimedean Banach algebra under the norm $\| \ \|$ given by $\|\mu\| = |\mu|(X)$, for all μ in $M(X)$. Further, the measures with compact support are norm dense in $M(X)$ and, if X is compact, $M(X)$ is an isometric linear copy of $C(X)^*$. (All these matters are proven in detail in [102].)

1.8 TOPOLOGICAL GROUPS AND SEMIGROUPS. Below we briefly describe the nature of groups and semigroups we will encounter in subsequent sections.

Let G be a commutative compact topological group. We define a *K-valued character of* G as a continuous homomorphism of G into $\{\alpha \in K : |\alpha| = 1\}$. Then all K-valued characters of G take their values in T_K if G is p-free. Below, we assume that G is zero-dimensional (so as to get the characters to separate the points of G).

By $o(G)$ we denote the set of all positive integers n for which G has an open subgroup H such that G/H contains an element of order n. We say G is p-*free* if $p \notin o(G)$. G is *compatible with* K if for every n in $o(G)$, T_K contains n elements of order n. If G is compatible with K, then G is p-free; further, one of the basic results in non-Archimedean harmonic analysis says that the K-valued characters of G separate points of G if and only if G is compatible with K ; see e.g. [102, page 360].

To avoid repetition we adopt the following convention:

Blanket Assumption. *For the rest of this chapter* S *denotes a commutative zero-dimensional topological semigroup with an identity element,* 1.

We say S is *torsional* if every compact subset of S is contained in a compact semigroup that is a union of groups. (In the case where S is a topological group, every compact subsemigroup of S is a compact group (by Theorem 1.2.4) and our definition of torsionality thus coincides with that employed in [102].)

If H is a compact subsemigroup of S and G a subgroup of H, then the closure of G is again a group. So a compact subsemigroup of S that is a union of groups is in fact a union of compact groups.

Evidently if S is torsional so is every closed subsemigroup of S.

If S is torsional, we say S is p-*free* if every compact subgroup of S is p-free. Further, S is said to be *compatible with* K if all compact subgroups of S are compatible with K.

2. SEMICHARACTERS

A *semicharacter* of S is a non-zero continuous homomorphism of S into the (discrete) semigroup $T_K \cup \{0\}$; we denote the set of all semicharacters of S by $S^\wedge$.

2.1 EXERCISE. $S^\wedge$ when endowed with the topology of uniform convergence on compacta is a zero-dimensional topological semigroup under pointwise multiplication.

Our main interest in this section is to determine when, for compact S, there are enough semicharacters to separate points of S. Towards this end we first collect together some observations on finite semigroups.

2.2. Let S be a finite commutative semigroup that is a union of groups. Then for every $x \in S$, $P(x) := \{x, x^2, x^3, \ldots\}$ is a subsemigroup of a finite group and hence is a group. Consequently P(x) contains a unique idempotent, which we denote by, e_x.

If x and y are any elements of S then $e_x e_y$ is an idempotent; choosing positive integers n and m such that $e_x = x^n$ and $e_y = x^m$, we have $e_x e_y = x^n y^m = (xy)^{nm}$. Hence $e_x e_y = e_{xy}$.

Finally, we note that the relation

$$e \leq e' \quad \text{if} \quad ee' = e$$

defines a partial ordering $\leq$ on E_S - the set of all idempotent elements of S.

2.3 EXERCISE.

(i) For all $f \in C(S)$ and $x \in S$ we have that ${}_x f, f_x \in C(S)$.

(ii) If U is a clopen subset of S then $\chi_{Ux^{-1}} \in C(S)$, for all $x \in S$.

Apart from using the following result in Theorem 2.6, it is also of independent interest.

2.4 PROPOSITION. [102, page 333] *Let* $\mathcal{U}$ *be a clopen partition of a compact semigroup* S; *then there exists a continuous homomorphism* π *of* S *into a finite semigroup* T, *such that the partition* $\{\pi^{-1}(t) : t \in T\}$ *refines* $\mathcal{U}$.

Proof. Let O(x) be the element of $\mathcal{U}$ containing x, for all x in S. We define an equivalence relation $\sim$ on S by

$$x \sim y \quad \text{if} \quad O(x) = O(y) \quad \text{and} \quad Ux^{-1} = Uy^{-1} \quad \text{for all} \quad U \in \mathcal{U}.$$

Since S is compact, $\mathcal{U}$ is finite. By Exercise 2.3(ii) each equivalence class of the relation $\sim$ is clopen and hence the classes form a finite clopen partition $\mathcal{V}$ of S.

For all x, y and z in S, if $x \sim y$, then $zx \sim zy$. Hence taking T to be the finite semigroup of all maps of $\mathcal{V}$ into itself, we have a well-defined map $\pi : S \to T$ given by

$$\pi(x)(\tilde{s}) := \widetilde{xs} \quad \text{for all} \quad s, x \in S;$$

where $\tilde{y}$ denotes the element of $\mathcal{V}$ containing y, for each y in S. The map π is clearly a homomorphism and, for all x in S,

$$\pi^{-1}(\pi(x)) = \{y \in S : \widetilde{ys} = \widetilde{xs} \quad \text{for all} \quad x \in S\}$$

$$= \bigcap_{s \in S} \{y \in S : O(ys) = O(xs) \text{ and } U(ys)^{-1} = U(xs)^{-1} \quad (U \in \mathcal{U})\}$$

- the latter is an intersection of closed sets, so every $\pi^{-1}(\pi(x))$ is closed, and π is a continuous homomorphism. Further, $\pi^{-1}\pi(x) \subseteq O(x)$, so $\{\pi^{-1}(t) : t \in T\}$ forms a refinement of $\mathcal{U}$.

2.5 LEMMA. *Let* S *be a finite semigroup that is a union of groups that are compatible with* K. *Then* S^ *separates points of* S.

Proof. Let a and b be any distinct points of S. Assuming the notation of item 2.2 we distinguish two cases:

Firstly, suppose $e_a = e_b$. For convenience, let $e := e_a$ and $G := \{x \in S : e_x = e\}$. Since $e_x e_y = e_{xy}$, for all x and y in S, it follows easily that G is a group and is compatible with K. We have $a,b \in G$ and so there exists a (K-valued) character, β, of G with $\beta(a) \neq \beta(b)$. We define a map $\gamma : S \to T_K \cup \{0\}$ by

$$\gamma(x) := \begin{cases} \beta(ex) & \text{if } e \leq e_x \\ 0 & \text{if } e \nleq e_x. \end{cases}$$

(Observe that, if $e \leq e_x$, then $e_{ex} = e_e e_x = ee_x = e$, so $ex \in G$.)

Then

$$\gamma(a) = \beta(ea) = \beta(a) \neq \beta(b) = \beta(eb) = \gamma(b).$$

Furthermore, if x and y are points of S with $e \leq e_{xy}$, then $e \leq e_x e_y$, so $e \leq e_x$ and $e \leq e_y$. It follows that γ is a semicharacter of S separating a and b.

Next, we suppose that $e_a \neq e_b$. We may assume that $e_a \leq e_b$. Consider the map $\gamma : S \to T_K \cup \{0\}$ given by

$$\gamma(x) := \begin{cases} 1 & \text{if } e_a \leq e_x \\ 0 & \text{if } e_a \nleq e_x. \end{cases}$$

Then $\gamma \in S^\wedge$, $\gamma(a) = 1$ and $\gamma(b) = 0$. This completes our proof.

2.6 THEOREM. *Let* S *be compact. Then the following items are equivalent:*

(i) $S^\wedge$ *separates points of* S;

(ii) S *is a union of compact groups that are compatible with* K; (*in other words,* S *is torsional and compatible with* K).

Proof. Suppose (i) holds. Then there exists a natural multiplicative and homeomorphic embedding of S into $H := (T_K \cup \{0\})^{S^\wedge}$. So we may assume that S is actually a compact subsemigroup of H. Since $T_K \cup \{0\}$ is a union of finite groups, all of them compatible with K, H is a union of compact groups, G_i, that are also compatible with K - say $H = \bigcup_{i \in I} G_i$. For every

$i \in I$ with $G_i \cap S$ nonempty, $G_i \cap S$ is a compact subsemigroup (hence a subgroup) of the group G_i compatible with K. Now $S = \bigcup_{i \in I} (G_i \cap S)$ and so (ii) follows.

Next we assume that (ii) holds and prove (i). Let a and b be any distinct elements of S. By Proposition 2.4, there exists a finite discrete semigroup T and a continuous homomorphism $\pi : S \to T$ such that $\pi(a) \neq \pi(b)$. By (ii), T is a union of groups that are compatible with K. By Lemma 2.5 we thus have a semicharacter γ in $T^\wedge$ such that $\gamma(\pi(a)) \neq \gamma(\pi(b))$. Hence $\gamma \circ \pi : S \to T_K \cup \{0\}$ is a semicharacter in $S^\wedge$ separating a and b. So item (i) follows. (Here $\gamma\circ\pi$ is given by $\gamma\circ\pi(x) := \gamma(\pi(x))$ $(x \in S)$.)

We would like to extend the following result to the semigroup case.

2.7 THEOREM (Schikhof [102, page 361]). *If* G *is a commutative, compact and zero-dimensional topological group that is compatible with* K, *then the set of all* (K-*valued*) *characters of* G *is an orthonormal base in* C(G).

Recalling that for a finite set X, *card*(X) stands for the number of elements in X, we first prove the following Lemma.

2.8 LEMMA. *Let* S *be finite, a union of groups and compatible with* K. *Then* $card(S^\wedge) = card(S)$ *and* $S^\wedge$ *is a linearly independent system in* K^S. (Note that $K^S = \ell^\infty(S)$ for finite S.)

Proof. By Lemma 2.5, $S^\wedge$ separates S. So Theorem 1.6 guarantees that the linear hull of $S^\wedge$ is K^S. Since $\dim(K^S) = card(S)$, to complete our proof it is sufficient to show that $card(S^\wedge) \leq card(S)$.

Recalling the notation of item 2.2, for each $e \in E_S$, let

$$G_e := \{x \in S : e_x = e\} \quad \text{and}$$

$$H_e := \{\gamma \in S^\wedge : e \leq e' \text{ for all } e' \in E_S \cap \gamma^{-1}(\{1\})\} .$$

Since S is a union of groups, one easily notes that $\{G_e\}_{e \in E_S}$ is a partition of S into groups and $\bigcup_{e \in E_S} H_e = S^\wedge$; so

$$card(S) = \sum_{e \in E_S} card(G_e) \quad \text{and} \quad card(S^\wedge) \leq \sum_{e \in E_S} card(H_e). \qquad (*)$$

By Theorem 2.7, we have card (G_e) = card$(G_e^\wedge)$ and so (*) suggests that the result will follow if we can show that card$(H_e) \leq$ card$(G_e^\wedge)$, for all $e \in E_S$. The latter will follow if we can prove that, for all $e \in E_S$, the map $\gamma \to \gamma|_{G_e}$ of H_e into $G_e^\wedge$ is injective.

Let $e \in E_S$ be fixed and suppose that $\beta|_{G_e} = \gamma|_{G_e}$ for some $\beta, \gamma \in H_e$. Taking x to be any fixed element of S, we now show that $\beta(x) = \gamma(x)$. If

$$e \leq e_x,$$

then $e_{ex} = e_e e_x = ee_x = e$, so $ex \in G_e$ and consequently

$$\beta(x) = \beta(e)\beta(x) = \beta(ex) = \gamma(ex) = \gamma(e)\gamma(x) = \gamma(x).$$

If, on the other hand,

$$e \leq e_x,$$

then $e_x \notin \beta^{-1}(\{1\})$, $\beta(e_x) \neq 1$, and so $\beta(e_x) = 0$. But then $e_x = x^n$ for some $n \in \mathbb{N}$, so $\beta(x) = 0$. Similarly $\gamma(x) = 0$. Hence $\beta(x) = \gamma(x)$. This completes our proof.

Now we can prove the promised extension of Theorem 2.7.

2.9 THEOREM. *Let* S *be compact, a union of groups and compatible with* K. *Then* $S^\wedge$ *is an orthonormal base of* C(S).

Proof. Theorem 2.6 says that $S^\wedge$ separates S and so, by Kaplansky's Theorem (see 1.6), the linear span of $S^\wedge$ is dense in C(S). So it remains to show that $S^\wedge$ is orthonormal. To this end, let $\gamma_1, \ldots, \gamma_n \in S^\wedge$ be distinct and $\lambda_1, \ldots, \lambda_n \in K$ be such that $\max\{|\lambda_i| : i = 1, \ldots, n\} = 1$. We need to show that $|\sum_{i=1}^{n} \lambda_i \gamma_i(x)| = 1$ for some $x \in S$.

Consider $\pi : S \to (T_K \cup \{0\})^n$ given by

$$\pi(x) := (\gamma_1(x), \ldots, \gamma_n(x)) \quad (x \in S).$$

Evidently $\pi(S)$ is a finite semigroup. (Recall that for $\lambda \in K$ with $|\lambda| \leq 1$ the symbol $\bar{\lambda}$ depicts the element of k corresponding to λ - see item 1.1)

We now consider the maps $\beta_i : \pi(S) \to k$ given by

$$\beta_i(\pi(x)) := \overline{\gamma_i(x)} \quad (x \in S \quad \text{and} \quad i \in \{1,\dots,n\}).$$

Each β_i is a k-valued semicharacter of $\pi(S)$ and $\beta_i \neq \beta_j$ for $i \neq j$. At least one $\bar{\lambda}_i$ is non-zero, so Lemma 2.8 applied to $\pi(S)$ and k in place of S and K, respectively, implies that $\sum_{i=1}^{n} \bar{\lambda}_i\beta_i \neq \bar{0}$. So there exists an x in S such that

$$\bar{0} \neq \sum_{i=1}^{n} \bar{\lambda}_i\beta_i(\pi(x)) = \sum_{i=1}^{n} \bar{\lambda}_i\bar{\gamma}_i(x) = \overline{\sum_{i=1}^{n} \lambda_i\gamma_i(x)}$$

- consequently $\left|\sum_{i=1}^{n} \lambda_i\gamma_i(x)\right| = 1.$

3. THE NON-ARCHIMEDEAN MEASURE ALGEBRA M(S)

Let H be any zero-dimensional topological semigroup. Then M(H) is a (non-Archimedean) Banach algebra with respect to the norm $\| \ \|$ and convolution multiplication $*$ given by

$$\nu*\mu(f) := \iint f(xy)d\nu(x)d\mu(y) = \iint f(xy)d\mu(y)d\nu(x)$$

for all $\nu,\mu \in M(H)$ and $f \in C(H)$.

Now for $\mu \in M(S)$, we define $\hat{\mu} \in \ell^{\infty}(S^\wedge)$ by

$$\hat{\mu}(\gamma) := \int \gamma \, d\mu \quad (\gamma \in S^\wedge)$$

and recall that the map $\mu \to \hat{\mu}$ is called the *Fourier-Stieltjes transform* of S. For all $\mu,\nu \in M(S)$ we have $(\mu*\nu)^\wedge = \hat{\mu}\hat{\nu}$.

3.1 THEOREM. *Let* S *be compact, a union of groups and compatible with* K. *Then the Fourier-Stieltjes transform of* S *is a Banach algebra isomorphism of* M(S) *onto* $\ell^{\infty}(S^\wedge)$.

Proof. Recalling Theorem 2.9 we have that $S^\wedge$ is an orthonormal base of C(S), so we have a linear isometry, π, of $c_o(S^\wedge)$ onto C(S) given by

$$\pi f := \sum_{\gamma \in S^\wedge} f(\gamma)\gamma \quad (\gamma \in c_o(S^\wedge) \text{ and } f \in c_o(S^\wedge))$$

Now $c_o(S^\wedge)^*$ is just $\ell^\infty(S^\wedge)$ and, since S is compact, M(S) can be identified with $C(S)^*$. Hence considering the adjoint, π^*, of π we have that, for all $\mu \in M(S)$ and $f \in c_o(S^\wedge)$,

$$(\pi^*\mu)(f) = \mu(\pi f) = \sum_{\gamma \in S} f(\gamma)\mu(\gamma) = \sum_{\gamma \in S^\wedge} f(\gamma)\hat{\mu}(\gamma) = \hat{\mu}(f).$$

Thus the Fourier-Stieltjes transform, $\mu \to \hat{\mu}$ $(\mu \in M(S))$, coincides with π^*. Since π^* is a linear isometry while $(\mu*\nu)^\wedge = \hat{\mu}\hat{\nu}$ $(\mu, \nu \in M(S))$, our Theorem follows.

Our next Theorem cites a situation where the Banach algebra M(S) is of the type hypothesised in Theorem 1.3.

3.2 THEOREM. *If* S *is torsional and* p-*free, then* M(S) *has power multiplicative norm.*

Proof. Let $\mu \in M(S)$ be fixed. By the remarks preceding Theorem 1.3 it is sufficient to show that $\|\mu*\mu\| = \|\mu\|^2$. Since tight measures with compact support are (norm-) dense in M(S), we may assume that $\text{supp}(\mu)$ is compact. By torsionality of S, $\text{supp}(\mu)$ is contained in some compact subsemigroup of S. So we may as well assume that S is compact.

Since K can be embedded into a complete algebraically closed field, we may also assume that S is compatible with K. Recalling Theorem 3.1, we then have

$$\|\mu*\mu\| = \|(\mu*\mu)^\wedge\| = \|\hat{\mu}^2\| = \|\hat{\mu}\|^2 = \|\mu\|^2.$$

Remark. It is interesting to note that even for finite S, if S is not a union of groups, M(S) may not have power multiplicative norm. For example the three element semigroup {0,1,a} in which 1 is an identity, 0 a zero and $a^2 = 0$, is such that $(\bar{a}-\bar{0})*(\bar{a}-\bar{0}) = 0$ - the zero-measure.

3.3. The reader can easily note that every compact subset T of S determines an equivalent relation, $\underset{T}{\sim}$, on S given by:

$\beta \underset{T}{\sim} \gamma$ *if and only if* $\beta = \gamma$ *on* T.

Recalling that $S^\wedge$ is endowed with the compact open topology we note that the equivalence classes $\mathcal{U}_T$, induced by $\underset{T}{\sim}$, form a clopen partition of $S^\wedge$. These partitions $\mathcal{U}_T$, with T ranging through all compact subsets of S, generate a non-Archimedean uniformity on $S^\wedge$. Let $UC(S^\wedge)$ denote the Banach algebra of all bounded functions $S^\wedge \to K$ that are uniformly continuous relative to this uniformity. We urge the reader to note that

(a) $UC(S^\wedge) \subseteq C(S^\wedge)$ and if S is a topological group then our definition of a uniformly continuous function on S coincides with that employed in [102].

(b) If S is torsional then for every clopen partition $\mathcal{U}$, that is uniform in the above sense, there exists a compact subsemigroup T of S such that $\mathcal{U}_T$ refines $\mathcal{U}$.

(c) Our notation gives the impression as if the uniformity is determined by the structure of $S^\wedge$. On the contrary, this is not the case, as shown in Item 5.2.

We wish to identify, for a large class of S, the range of the Fourier-Stieltjes transform. Towards this end we have the following Proposition.

3.4 PROPOSITION. *Let* S *be torsional and* $S^\wedge$ *separate the points of* S. *Then the Fourier-Stieltjes transform maps* M(S) *isometrically into* $UC(S^\wedge)$.

Proof. Let T be any compact subsemigroup of S that contains 1. We first show that semicharacters of T extend to semicharacters of S. To the latter end, let R_T be the restriction map $S^\wedge \to T^\wedge$. As $T^\wedge$ separate the points of T, Theorem 2.6 says that T is compatible with K; so $T^\wedge$ is an orthonormal base of C(T), (by Theorem 2.9), containing $R_T(S^\wedge)$. Thus if β is any element of $T^\wedge$, either $\beta \in R_T(S^\wedge)$ or $B \perp \overline{\text{span}}\, R_T(S^\wedge)$. Since $R_T(S^\wedge)$ separates T we must have $C(T) = \overline{\text{span}}\, R_T(S^\wedge)$, by Theorem 1.6 and so $\beta \in R_T(S^\wedge)$. Thus $T^\wedge = R_T(S^\wedge)$.

Now fix $\mu \in M(S)$ with compact support. We can find a compact subsemigroup T with $1 \in T$ and $\text{supp}(\mu) \subseteq T$. As above, T is compatible with K and so

$$\|\mu\| = \sup\{|\int \beta d\mu| : \beta \in T^\wedge\} \qquad \text{by Theorem 3.1}$$

$$= \sup\{|\int (\gamma_{|T}) d\mu| : \gamma \in S^\wedge\} \qquad \text{since } T^\wedge = R_T(S^\wedge)$$

$$= \sup\{|\hat{\mu}(\gamma)| : \gamma \in S^\wedge\} \qquad \text{since } \mathrm{supp}(\mu) \subseteq T$$

$$= \|\hat{\mu}\| .$$

Evidently $\hat{\mu}$ is constant on the elements of the partition $\mathcal{U}_T$ and so $\hat{\mu} \in UC(S^\wedge)$. Since measures with compact supports are dense in M(S), our result follows.

We now present our desired target by simply sharpening the conclusion of our preceding Proposition. For the case of a topological group see e.g. [102, Theorem 9.20].

3.5 THEOREM. *Let* S *be torsional and* $S^\wedge$ *separate the points of* S. *Then the Fourier-Stieltjes transform is a Banach algebra isomorphism of* M(S) *onto* $UC(S^\wedge)$.

Proof. In view of Proposition 3.4 it is sufficient to show that $A := \{f \in UC(S^\wedge) : f = \hat{\mu} \text{ for some } \mu \in M(S)\}$ is dense in $UC(S^\wedge)$. To this end let $g \in UC(S^\wedge)$ and $\varepsilon > 0$ be given. Recalling Remark 3.3(b), we can find a compact subsemigroup T of S containing 1 such that

$$|g(\beta) - g(\gamma)| \leq \varepsilon \quad \text{for all} \quad \beta, \gamma \in S^\wedge \quad \text{with } \beta_{|T} = \gamma_{|T} .$$

Corresponding to each element I_α of the partition $\mathcal{U}_T$ choose one and only one $\gamma_\alpha \in I_\alpha$ and consider the function $f : S^\wedge \to K$ given by

$$f(\beta) = g(\gamma_\alpha) \quad \text{if and only if} \quad \beta \in I_\alpha, \quad \text{for each } I_\alpha \in \mathcal{U}_T.$$

Then f is constant on elements of $\mathcal{U}_T$ and $|g(\gamma) - f(\gamma)| \leq \varepsilon$ for all $\gamma \in S^\wedge$. So our result will follow if we show that $f \in A$.

As noted in the proof of Proposition 3.4 every element of $T^\wedge$ extends to an element of $S^\wedge$ and so we may define $h \in \ell^\infty(T^\wedge)$ by

$$h(\gamma_{|T}) := f(\gamma) \qquad (\gamma \in S^\wedge).$$

By Theorem 3.1, we can find $\nu \in M(T)$ such that $h(\beta) = \int_T \beta d\nu$ for all $\beta \in T^\wedge$. Defining $\mu \in M(S)$ by $\mu(E) := \nu(E \cap T)$ for all clopen $E \subseteq S$, we have

$$f(\gamma) = h(\gamma_{|T}) = \int \gamma_{|T}\, d\nu = \int \gamma d\mu = \hat{\mu}(\gamma) \qquad (\gamma \in S^\wedge).$$

Thus $f = \hat{\mu}$ and $f \in A$. As remarked above, this completes our proof.

4. THE NON-ARCHIMEDEAN SEMIGROUP ALGEBRA $M_n(S)$

Let H be any zero-dimensional topological semigroup. For $x \in H$ we define $\bar{x} \in M(H)$ by

$$\bar{x}(U) := \begin{cases} 1 & \text{if } x \in U \\ 0 & \text{if } x \notin U, \quad \text{for all } U \in \mathcal{B}(H). \end{cases}$$

(With the context being different we do not expect the reader to confuse this notation with that in item 1.1.)

Let

$$M_n^\ell(H) := \{\mu \in M(H) : \text{the map } x \to \bar{x}*\mu \text{ of } H \text{ into } M(H) \text{ is norm continuous}\}$$

$$M_n^r(H) := \{\mu \in M(H) : \text{the map } x \to \mu*\bar{x} \text{ of } H \text{ into } M(H) \text{ is norm continuous}\}$$

$$M_n(H) := M_n^\ell(H) \cap M_n^r(H).$$

For a zero-dimensional topological group G, Schikhof [102, Theorem 8.9] showed that if $M_n^\ell(G) \neq \{0\}$ then G is locally compact and carries a K-valued Haar measure λ such that $M_n^\ell(G) = M_n(G) = L(G) := \{\mu \in M(G) : \mu = j\lambda$ for some continuous $j : G \to K\}$ - hence the reason for our interest in sets like $M_n(H)$.

First we make some remarks on integration of vector valued functions. For all $\nu \in M(H)$, all Banach spaces E and all bounded continuous functions $f : H \to E$, in a reasonable way one can define an integral $\int f(x)d\nu(x)$ that is an element of E. If E and F are Banach spaces and $T : E \to F$ is a linear continuous map, then $\int T(f(x))d\nu(x) = T(\int f(x)d\nu(x))$. In particular noting that $M_n^\ell(H)$ is clearly a Banach space, if $E := M_n^\ell(H)$ and $f(x) := \bar{x}*\mu$ for some $\mu \in M_n^\ell(H)$, since the maps $y \to \bar{y}*(\bar{x}*\mu) = \overline{yx}*\mu$ of H into M(H) are clearly continuous $(x \in H)$ f is thus a continuous function of H into $M_n^\ell(H)$

and so we obtain an element $\int \bar{x}*\mu d\nu(x)$ of $M_n^{\ell}(H)$. For any clopen $A \subseteq H$,

$$(\int \bar{x}*\mu\, d\nu(x))(A) = \int \bar{x}*\mu(A)d\nu(x) = \nu*\mu(A),$$

so $\nu*\mu = \int \bar{x}*\mu\, d\nu(x)$. Thus $\nu*\mu \in M_n^{\ell}(H)$. More generally we have the following result.

4.1 THEOREM. *For any zero-dimensional topological semigroup* H, *we have that* $M_n^{\ell}(H)$ *and* $M_n(H)$ *are closed ideals of* $M(H)$.

Proof. It is sufficient to prove the result for $M_n^{\ell}(H)$. Evidently $M_n^{\ell}(H)$ is closed. Let $\mu \in M_n^{\ell}(H)$ and $\nu \in M(H)$ be fixed. Then as observed above $\nu*\mu \in M_n^{\ell}(H)$.

For all $x,y \in H$ we have

$$||\bar{x}*(\mu*\nu) - \bar{y}*(\mu*\nu)|| \leq ||\bar{x}*\mu - \bar{y}*\mu||\ ||\nu||$$

and so $\mu*\nu \in M_n^{\ell}(H)$. This completes the proof.

Taking $S_n^{\wedge} := \{\gamma \in S^{\wedge}: \hat{\mu}(\gamma) \neq 0$ for some $\mu \in M_n(S)\}$ and $M_n(S)^{\wedge}$ to be the set of all non-zero multiplicative linear functionals on $M_n(S)$, we have the following non-Archimedean analogue of Theorem 2.5.6.

4.2 THEOREM. $S_n^{\wedge}$ *with the compact-open topology is homeomorphic to* $M_n(S)^{\wedge}$ *with the Gel'fand topology.*

Proof. Let $h \in M_n(S)^{\wedge}$ be fixed and choose $\mu \in M_n(S)$ such that $h(\mu) \neq 0$. Consider $\gamma_h : S \to K$ given by

$$\gamma_h(x) := \frac{h(\bar{x}*\mu)}{h(\mu)} \quad .$$

Then, as in the proof of Theorem 2.5.3, one can easily show that $\gamma_h \in S_n^{\wedge}$ and that each $\gamma \in S_n^{\wedge}$ arises from a unique $h \in M_n(S)$ with $\gamma_h = \gamma$; thus the map $h \to \gamma_h$ of $M_n(S)^{\wedge}$ into $S_n^{\wedge}$ is both one-to-one and onto.

We now show that the map $h \to \gamma_h$ is a homeomorphism in a similar but simpler way to the proof of Theorem 2.5.6.

Suppose $h_\alpha \to h$ in $M_n(S)^{\wedge}$ and let $T \subseteq S$ be compact. We wish to show that $\gamma_{h_\alpha} \to \gamma_h$ uniformly on T. We can choose $\mu \in M_n(S)$ with $h(\mu) = 1$ and may assume that, for all α, we have $|h_\alpha(\mu) - h(\mu)| < 1$. Then

$$1 = |(h_\alpha(\mu)-h(\mu))-h_\alpha(\mu)| \leq \max\{|h_\alpha(\mu)-h(\mu)|, |h_\alpha(\mu)|\} = |h_\alpha(\mu)|$$

and

$$|h_\alpha(\mu)| = |(h_\alpha(\mu)-h(\mu))+h(\mu)| \leq \max\{|h_\alpha(\mu)-h(\mu)|,|h(\mu)|\} = |h(\mu)|,$$

so $|h_\alpha(\mu)| = 1$. Now the elements $h_\alpha(\mu)^{-1}h_\alpha$ of $M_n(S)^*$ form a norm bounded net that converges in the weak*-topology; so it converges uniformly on the compact subsets of $M_n(S)$. As $\{\bar{x}*\mu : x \in T\}$ is a compact subset, we have that

$$h_\alpha(\mu)^{-1}h_\alpha(\bar{x}*\mu) \to h(\mu)^{-1}h(\bar{x}*\mu) \quad \text{uniformly for } x \in T.$$

Thus $\gamma_{h_\alpha} \to \gamma_h$ uniformly on T.

Next, suppose $\gamma_\alpha \to \gamma$ in $S_n^\wedge$. For each $\nu \in M_n(S)$ and $\varepsilon > 0$ we can choose $\rho \in M(S)$ such that $T := \text{supp}(\rho)$ is compact and $\|\nu-\rho\| \leq \varepsilon$. For large enough α, we have $\gamma_\alpha = \gamma$ on T, so $\int \gamma_\alpha d\rho = \int \gamma\, d\rho$ and (with $\gamma_\alpha = \gamma_{h_\alpha}$ and $\gamma = \gamma_h$)

$$|\hat{\nu}(h_\alpha) - \hat{\nu}(h)| = |\int(\gamma_\alpha-\gamma)d\nu| = |\int(\gamma_\alpha-\gamma)d(\nu-\rho)| \leq \varepsilon.$$

Hence $h_\alpha \to h$ in $M_n(S)^\wedge$.

4.3 THEOREM. *Let* S *be torsional and compatible with* K. *Then the canonical map of* $M_n(S)$ *into* $\ell^\infty(S^\wedge)$ *is an isometric algebra homomorphism.*

Proof. First we note that the map $\mu \to \hat{\mu}$ of $M_n(S)$ into $\ell^\infty(S^\wedge)$ has the properties

$$(\nu*\mu)^\wedge = \hat{\nu}\hat{\mu} \quad\text{and}\quad \|\hat{\mu}\| \leq \|\mu\| \qquad (\nu,\mu \in M_n(S)).$$

Since S is compatible with K, it is p-free and so M(S) has power multiplicative norm, by Theorem 3.2. So for every $\mu \in M_n(S)$ a result of Springer (see Theorem 1.3) implies the existence of a multiplicative linear map $\phi : M(S) \to L$, where L is some complete field containing K, such that

$$|\phi(\mu)| = \|\mu\| \quad\text{and}\quad \|\phi\| = 1.$$

Setting $\gamma(x) := \phi(\bar{x})$ we have a bounded homomorphism $\gamma : S \to L$. Now $\gamma(x)\phi(\mu) = \phi(\bar{x})\phi(\mu) = \phi(\bar{x}*\mu)$ $(x \in S)$, so γ is continuous. Since S is compatible with K, it follows that $\gamma(x) \in K$ for all $x \in S$ and $\gamma \in S^\wedge$.

Recalling the remarks preceding Theorem 4.1, for any $\nu \in M(S)$, we have

$$\phi(\nu)\phi(\mu) = \phi(\nu*\mu) = \phi\left(\int \bar{x}*\mu d\nu(x)\right) = \int \phi(\bar{x}*\mu)d\nu(x)$$

$$= \int \gamma(x)\phi(\mu)d\nu(x) = \hat{\nu}(\gamma)\phi(\mu).$$

So $\phi(\nu) = \hat{\nu}(\gamma)$, in particular $\phi(\mu) = \hat{\mu}(\gamma)$. Hence

$$||\mu|| = |\phi(\mu)| = |\hat{\mu}(\gamma)| \leq ||\hat{\mu}||$$

and so $||\mu|| = ||\hat{\mu}||$. This completes our proof.

A comparison of Theorems 4.3 and 3.5 motivates one to seek for the image of $M_n(S)$ under the Fourier-Stieltjes transform. Towards this end we first give a definition.

4.4 DEFINITION. A continuous function $f : S^\wedge \to K$ is said to be *arbitrarily small outside equicontinuous sets* if given $\varepsilon > 0$ we can find an equicontinuous set $A \subset S^\wedge$ such that $|f| \leq \varepsilon$ on $S^\wedge \setminus A$; the set of all such f is denoted by $EQ_o(S^\wedge)$.

The following lemma is also of independent interest.

4.4 LEMMA. *We have the following inclusion relations.*

(i) $EQ_o(S^\wedge) \subseteq C_o(S^\wedge)$.

(ii) $C_o(S^\wedge) \subseteq UC(S^\wedge)$.

Proof. (i) Let $f \in EQ_o(S^\wedge)$ be fixed. Given $\varepsilon > 0$, we consider an equivalence relation $\sim$ on S defined by

$$x \sim y \text{ if } \gamma(x) = \gamma(y) \text{ for all } \gamma \in S^\wedge \text{ with } |f(\gamma)| \geq \varepsilon.$$

The equivalence classes are evidently clopen and hence the quotient semigroup $R := S/_\sim$ is discrete. Let $\pi : S \to R$ be the quotient map and $\hat{\pi} : R^\wedge \to S^\wedge$ the dual map of π. Since $R^\wedge$ is compact (as R is discrete) and $\hat{\pi}$ is continuous, we have that $\hat{\pi}(R^\wedge)$ is a compact subset of $S^\wedge$. By our definition of $\sim$ we have $|f| < \varepsilon$ outside the compact set $\hat{\pi}(R^{\hat{}})$ and so $f \in C_o(S^{\hat{}})$.

(ii) It is sufficient to show that a characteristic function of a compact and open subset of $S^{\wedge}$ is in $UC(S^{\wedge})$. Let $V \subset S^{\wedge}$ be compact and open. Recalling the definition of compact open topology on $S^{\wedge}$, for each $\beta \in V$ we can find a compact $K_\beta \subset S$ such that

$$O(\beta) := \{\gamma \in S^{\wedge} : \beta = \gamma \text{ on } K_\beta\} \subseteq V$$

(and $O(\beta)$ is a neighbourhood of β). So for some $n \in \mathbb{N}$, we have $\beta_1,\ldots,\beta_n \in V$ such that $V = O(\beta_1)\cup\ldots\cup O(\beta_n)$. Setting $T := K_{\beta_1}\cup\ldots\cup K_{\beta_n}$ we see that V is a union of elements of the clopen partition $\mathcal{U}_T$ (defined in item 3.3) and so the characteristic function of V lies in $UC(S^{\wedge})$.

4.5 THEOREM. *Let* S *be torsional and* $S^{\wedge}$ *separate points of* S. *Then*

(i) *The Fourier-Stieltjes transform is a Banach algebra isomorphism of* $M_n(S)$ *onto* $EQ_o(S^{\wedge})$.

(ii) *If every compact subset of* $S^{\wedge}$ *is equicontinuous, then the Fourier-Stieltjes transform is a Banach algebra isomorphism of* $M_n(S)$ *onto* $C_o(S^{\wedge})$.

Proof. (i) Given $\mu \in M_n(S)$ and $\varepsilon > 0$, we consider the set $A := \{\gamma \in S^{\wedge} : |\hat{\mu}(\gamma)| \geq \varepsilon\}$. For all $x,y \in S$

$$\|\bar{x}*\mu - \bar{y}*\mu\| \geq \sup_{\gamma\in A} |(\bar{x}*\mu - \bar{y}*\mu)^{\wedge}(\gamma)| = \sup_{\gamma\in A} |\gamma(x) - \gamma(y)||\hat{\mu}(\gamma)|$$

$$\geq \varepsilon \sup_{\gamma\in A} |\gamma(x) - \gamma(y)|;$$

hence A is equicontinuous.

By Lemma 4.4(i) and Theorem 3.5 we have that the map $\mu \to \hat{\mu}$ is an isometric algebra homomorphism of $M_n(S)$ into $EQ_o(S^{\wedge})$. To show that this map is onto, fix $f \in EQ_o(S^{\wedge})$ and choose $\nu \in M(S)$ such that $f = \hat{\nu}$ (which can be done, by Theorem 3.5 and Lemma 4.4). We now show that $\nu \in M_n(S)$.

Fix $x \in S$ and $\varepsilon > 0$. By our choice of f the set $A := \{\gamma \in S^{\wedge} : |\hat{\nu}(\gamma)| \geq \varepsilon\}$ is equicontinuous, so $O(x) := \{y \in S : |\gamma(x) - \gamma(y)| < \|\nu\|^{-1}\varepsilon \text{ for all } \gamma \in A\}$ is an open neighbourhood of x. Now, for all $y \in O(x)$ we have

$$(*) \qquad |\gamma(x) - \gamma(y)||\hat{\nu}(\gamma)| \begin{cases} \leq \|\nu\|^{-1} \varepsilon \ \|\nu\| = \varepsilon & \text{if } \gamma \in A \\ \leq 1.\varepsilon = \varepsilon & \text{if } \gamma \in S\hat{}\setminus A. \end{cases}$$

Recalling Theorem 3.5, for all $y \in O(x)$, we have

$$\begin{aligned} \|\bar{x}*\nu - \bar{y}*\nu\| = \|(\bar{x}*\nu - \bar{y}*\nu)\hat{}\| &= \sup_{\gamma \in S\hat{}} |(\bar{x}*\nu - \bar{y}*\nu)\hat{}(\gamma)| \\ &= \sup_{\gamma \in S\hat{}} |\gamma(x) - \gamma(y)||\hat{\nu}(\gamma)| \\ &\leq \varepsilon . \qquad \text{by } (*). \end{aligned}$$

Consequently $\nu \in M_n(S)$ and item (i) is proven.

Item (ii) now follows from item (i) and Lemma 4.4(i) together with the fact that now $C_o(S\hat{}) \subseteq EQ_o(S\hat{})$.

4.6 REMARKS. Even in the case of topological groups, the requirement that every compact subset of $S\hat{}$ be equicontinuous employed in Theorem 4.5(ii) is not redundant. We refer the reader to [102, page 356] for an example : There, a torsional zero-dimensional topological group G is constructed such that $G\hat{}$ is compact but G is not locally compact. Thus $M_n(G) = \{0\}$ while $C_o(G\hat{}) = C(G\hat{}) \neq \{0\}$.

Nevertheless the class of topological semigroups S for which compact subsets of $S\hat{}$ are equicontinuous is large. Below we cite some of such cases:

(a) *Suppose* S (*is torsional and*) *has the property : for every subset* O *of* S *with* $O \cap T$ *clopen in* T *for all compact* $T \subseteq S$ *we have that* O *is open.* (This property is met, for example, if S is locally compact or metrizable and compares closely with the definition of a k-space.) *Then every compact subset* A *of* $S\hat{}$ *is equicontinuous.*

Proof. Let A be a compact subset of $S\hat{}$. Our result will follow if we show that $O(x) := \{y \in S : \gamma(y) = \gamma(x) \text{ for all } \gamma \in A\}$ is clopen $(x \in S)$. To this end, let T be a compact subsemigroup of S with $1 \in T$ and note that the restriction map $S\hat{} \to T\hat{}$ is continuous and $T\hat{}$ is discrete. Hence

$A_1 := \{\gamma_{|T} : \gamma \in A\}$ is (compact and so) finite. Now

$$O(x) \cap T = \bigcap_{\gamma \in A} \{y \in T : \gamma(y) = \gamma(x)\} = \bigcap_{\gamma \in A_1} \{y \in T : \gamma(y) = \gamma(x)\},$$

so $O(x) \cap T$ is clopen in T. By the given property of S, O(x) is clopen.

(b) *Suppose* S (*is torsional and*) *has the property: for every* $x \in S$ *and* $\gamma \in S^\wedge$ *there is a neighbourhood* U *of* x *and an* $a \in S$ *such that* Ua *is relatively compact and* $\gamma(a) \neq 0$. *Then compact subsets of* $S^\wedge$ *are equicontinuous.*

Proof. Let $A \subseteq S^\wedge$ be compact and $x \in S$ be fixed. We show that O(x), defined as in case (a), is a neighbourhood of x.

For every $\gamma \in A$, let U_γ be a neighbourhood of x and a_γ an element of S such that $U_\gamma a_\gamma$ is relatively compact and $\gamma(a_\gamma) \neq 0$. By compactness of A, there exists finitely many points $\gamma_1,\ldots,\gamma_n$ in A such that

$$A \subseteq \bigcup_{i=1}^{n} \{\beta \in S^\wedge : \beta(a_{\gamma_i}) \neq 0\}.$$

Let $U := U_{\gamma_1} \cap \ldots \cap U_{\gamma_n}$ and, for convenience, write a_i in place of a_{γ_i} and note that Ua_i is relatively compact, for $i = 1,\ldots,n$.

Since S is torsional, there is a compact subsemigroup T of S containing the (compact) set $\{1,a_1,\ldots,a_n\} \cup (U\{a_1,\ldots,a_n\})$. As noted in the proof of (a), $\{\gamma_{|T} : \gamma \in A\}$ is finite, so there exist finitely many points $\beta_1,\ldots,\beta_m \in A$ such that

for each $\beta \in A$, $\beta_{|T} = \beta_{j|T}$ for some $j \in \{1,\ldots,m\}$.

Now $W := \{y \in U : \beta_j(y) = \beta_j(x)$ for $j = 1,\ldots,m\}$ is indeed a neighbourhood of x, so A will be equicontinuous at x if we can show that $W \subseteq O(x)$.

To the latter end, let $y \in W$ and $\beta \in A$. There exist $j \in \{1,\ldots,m\}$ and $i \in \{1,\ldots,n\}$ such that

$$\beta_{|T} = \beta_{j|T} \quad \text{and} \quad \beta(a_i) \neq 0.$$

Since

$ya_i \in Wa_i \subseteq ua_i \subseteq T$ and $a_i \in T$,

$\beta(y)\beta(a_i) = \beta(ya_i) = \beta_j(ya_i) = \beta_j(y)\beta_j(a_i) = \beta_j(y)\beta(a_i)$ and so $\beta(y) = \beta_j(y)$. In particular $\beta(x) = \beta_j(x)$, whence $\beta(y) = \beta(x)$ and $y \in O(x)$. Thus $W \subset O(x)$ and we are done.

In general, the properties of S stated in (a) and (b) are distinct - in fact neither implies the other; for counter-examples, we refer the interested reader to [40; item 4.8]. *It remains an open problem whether one can obtain a general characterization of those semigroups* S *for which compact subsets of* S^ *are equicontinuous.*

5. NOTES

5.1. It would be interesting if one could discover some meaningful non-Archimedean analogues of the spaces $M_a(S)$ and $M_e(S)$ studied in Chapter 2. As noted above the object $M_n(S)$ seems to be quite interesting and a lot remains to be studied - for instance, it is not clear what the "foundation" of such an object is. We however suspect that it could be reasonable to define the foundation of $A \subseteq M(S)$ as the smallest closed set containing the set $\{x \in S : N_\mu(x) \neq 0$ for some $\mu \in A\}$.

5.2. The structure of S^ does not determine the uniformity defined in item 3.3. In fact *one can find zero-dimensional commutative topological semigroups* S *and* T *with identity elements and such that:* S^ *and* T^ *are identical as topological semigroups but* UC(S^) ≠ UC(T^). See e.g. [40, Example 4.4].

5.3. A lot needs to be investigated on S . For instance *if* S *is locally compact, torsional and compatible with* K, *we conjecture that* S^ *separates points of* S. See also [40, Section 4] for some examples on the structure of S^.

5.4. As noted at the beginning of this chapter, the results studied here are taken from the paper of Dzinotyiweyi and van Rooij [40]. The reader who wishes to research more on abstract harmonic analysis of this type may find it helpful to study related results in the group case. References for the latter include the theses of H.W. Schikhof, "*Non-Archimedean harmonic analysis*", Katholieke Universiteit, Nijmegen (1967); A.M.M. Gommers [47] and L. Duponcheel, "*Non-Archimedean induced representations and related topics*", Vrije Universiteit Brussel (1979).

Appendix A The strict topology and duality theorem

Throughout this Appendix X denotes a C-distinguished topological space.

1. THE STRICT TOPOLOGY

A.1. A non-negative function ψ in m(X) is called a *hood* if the set $\{x \in X : \psi(x) \geq \varepsilon\}$ is compact, for every $\varepsilon > 0$. Let Hd(X) be the set of all hoods on X. The *strict topology* (or equivalently, the β-*topology*) on C(X) is the topology generated by the seminorms $\{p_\psi : \psi \in Hd(X)\}$, where

$$p_\psi(f) := \|\psi f\|_X = \sup\{\psi(x)|f(x)| : x \in X\}$$

for all f in C(X) and ψ in Hd(X). We write $C_\beta(X)$ for C(X) endowed with the strict topology. (See e.g. [12], [61] and [103] for more details on the strict topology.)

2. A GENERALIZED STONE-WEIERSTRASS THEOREM

A.2. THEOREM (Giles [43]). *Let* A *be a* β-*closed subalgebra of* C(X) *which separates points and contains, for each* x *in* X, *a function nonzero at* x. *Then* A = C(X).

We are particularly interested in a Corollary of this Theorem (given below). Towards the latter end we prove a lemma.

A.3. LEMMA. *Let* μ *be a positive measure in* M(X) *and* F *a compact subset of* X. *Then*

(i) $\mu(F) = \mathit{inf}\{\mu(f) : f \in C(X)$ is a positive function with $\chi_F \leq f\}$

(ii) C(X) *is a dense subspace of* $L^1(X,\mu)$.

Proof. (i) Let $\varepsilon > 0$ be given and choose an open neighbourhood U of F such that $\mu(U) < \mu(F) + \frac{1}{2}\varepsilon$. We can find an increasing sequence of compact sets $\{C_n\} \subseteq X \setminus U$ such that $\mu(\bigcup_{n=1}^{\infty} C_n) = \mu(X \setminus U)$. So we can choose $n_o \in \mathbb{N}$ such that

$$n \geq n_o \text{ implies that } \mu(X \setminus U) < \mu(C_n) + \tfrac{1}{2}\varepsilon.$$

Since F and C_n are disjoint compact sets and X is C-distinguished, a standard compactness argument shows that there is a real-valued function $f_n \in C(X)$ such that

$$f_n(X) \subseteq [0,1], \quad f_n(C_n) = \{0\} \text{ and } f_n(F) = \{1\}.$$

Hence, for $n \geq n_o$, we get

$$\begin{aligned} \mu(F) \leq \int f_n(x)d\mu(x) &= \int_U f_n(x)d\mu(x) + \int_{X\setminus U} f_n(x)d\mu(x) \\ &\leq \mu(U) + \mu((X\setminus U)\setminus C_n) \\ &< \mu(F) + \varepsilon. \end{aligned}$$

This completes our proof for item (i).

(ii) Let A be any Borel subset of X. We first show that χ_A is a cluster point of C(X) in $L^1(X,\mu)$. To this end, let $\varepsilon > 0$ be given. There is a compact set F and an open set U such that $F \subseteq A \subseteq U$ and $\mu(U\setminus F) < \frac{1}{4}\varepsilon$. Now let the sequence of compact sets $\{C_n\}$ and of functions $\{f_n\}$ be chosen as in (i) with respect to F and U. We then get

$$\begin{aligned} \|\chi_A - f_n\|_1 &:= \int |\chi_A - f_n|(x)d\mu(x) = \int_U |\chi_A - f_n|(x)d\mu(x) + \int_{X\setminus U} |\chi_A - f_n|(x)d\mu(x) \\ &= \int_{U\setminus F} |\chi_A - f_n|(x)d\mu(x) + \int_{(X\setminus U)\setminus C_n} f_n(x)d\mu(x) \\ &\leq 2\mu(U\setminus F) + \mu((X\setminus U)\setminus C_n) \\ &< \tfrac{1}{2}\varepsilon + \tfrac{1}{2}\varepsilon = \varepsilon \quad \text{for } n \geq n_o. \end{aligned}$$

It is now evident from the first paragraph of the proof of Hewitt and Ross [59, Theorem 12.10] that item (ii) holds.

A.4. COROLLARY. *Let A be a subalgebra of* C(X) *such that A separates points and for each* x *in* X *some member of A does not vanish at* x. *Then A is dense in* $L^1(X,\mu)$ *for every positive measure* μ *in* M(X).

Proof. Let μ be a given positive measure in $M(X)$. Suppose $f \in L^1(X,\mu)$ and $\varepsilon > 0$ are given. By item (ii) of our Lemma, we can find $g \in C(X)$ with $\int |f-g|(x)d\mu(x) < \frac{1}{2}\varepsilon$.

There is an increasing sequence $\{K_n\}$ of compact subsets such that $\mu(\bigcup_{n=1}^{\infty} K_n) = \|\mu\|$. Setting $K_o := \phi$ and $a_n := \mu(K_n \setminus K_{n-1})$ we have that the function $\psi := \sum_{n=1}^{\infty} a_n \chi_{K_n \setminus K_{n-1}}$ is a hood. Then, by Theorem A.2,

$$p_\psi(g-h) < \tfrac{1}{2}\varepsilon \quad \text{for some} \quad h \in A.$$

Hence

$$\begin{aligned}\int |f-h|(x)d\mu(x) &\leq \int |f-g|(x)d\mu(x) + \int |g-h|(x)d\mu(x) \\ &< \tfrac{1}{2}\varepsilon + \sum_{n=1}^{\infty} \int_{K_n \setminus K_{n-1}} |g-h|(x)d\mu(x) \\ &\leq \tfrac{1}{2}\varepsilon + \sum_{n=1}^{\infty} a_n \|g-h\|_{K_n \setminus K_{n-1}} = \tfrac{1}{2}\varepsilon + p_\psi(g-h) < \varepsilon.\end{aligned}$$

3. A GENERALIZATION OF THE RIESZ-REPRESENTATION THEOREM

The following Theorem is due to Giles [43] for the case where X is completely regular. The present form was noted by Hirschfeld [61].

A.5. THEOREM. *For each* μ *in* $M(X)$ *let* h_μ *be the functional on* $C(X)$ *given by* $h_\mu(f) := \int f(x)d\mu(x)$ $(f \in C(X))$. *Then the map* $\mu \to h_\mu$ *defines a linear isomorphism from* $M(X)$ *onto* $C_\beta(X)^*$ - (the dual of $C_\beta(X)$).

A.6. COROLLARY. We have that

(i) $\|\mu\| = \sup\{|\mu(f)| : f \in C(X)$ and $\|f\|_X \leq 1\}$ *for all* μ *in* $M(X)$.

(ii) $M(X)$, *under the norm* $\| \ \|$, *is a Banach space. Equivalently* $C_\beta(X)^*$ *is a closed subspace of* $C(X)^*$.)

Proof. Item (i) follows trivially from Lemma A.3(ii). Now $M(X)$ is clearly a (complex) normed linear space, so to prove item (ii) it remains to show that $M(X)$ is complete. To this end, suppose $\{\mu_n\}$ is a Cauchy sequence in $M(X)$. Then item (i) implies that $\{\mu_n\}$ is a Cauchy sequence in the Banach space $C(X)^*$ and so converges to some μ in $C(X)^*$.

We now show that μ is strictly continuous. Let $\varepsilon > 0$ be given and choose n_o such that $\|\mu_{n_o} - \mu\|_{C(X)^*} < \varepsilon$. Since μ_{n_o} is strictly continuous, there is a hood ψ such that if

$$p_\psi(f) < 1 \quad \text{then} \quad |\mu_{n_o}(f)| < \varepsilon \qquad (f \in C(X)).$$

Hence, for such f, we have

$$|\mu(f)| < |\mu_{n_o}(f)| + \varepsilon < \varepsilon(\|f\|_X + 1).$$

Thus μ is strictly continuous. By Theorem A.4 we must have μ in M(X) and our result follows.

Appendix B Weakly compact sets

1. WEAKLY COMPACT SUBSETS OF A BANACH SPACE

B.1. THEOREM (Eberlein-Smulian). *Let* C *be a subset of a Banach space* A. *Then the following items are equivalent:*

(i) C *is relatively weakly compact;*

(ii) *each sequence in* C *has a weakly convergent subsequence in* A (i.e. C *is sequentially weakly compact*).

The reader can find the proof of the preceding Theorem in e.g. Dunford and Schwartz [29, V.6.1]. A proof for the following Theorem can be found in Glicksberg [46].

B.2. THEOREM (Krein-Smulian). *If* C *is a weakly compact subset of a Banach space* A *then the closure of the convex hull of* C *is weakly compact.*

The next Theorem is proved by Kelley and Namioka [66, page 160].

B.3. THEOREM. *The weak topology for a product of linear topological spaces is the product of the weak topologies for the coordinate spaces.*

2. WEAKLY COMPACT SUBSETS OF M(X)

The equivalence of the first two items of our next Theorem is due to Grothendieck [55, Theorem 2] and that of the last two items can be deduced from Stein [99, pages 139 and 140]. (In fact the whole Theorem is essentially due to Grothendieck [55].)

B.4. THEOREM. *Let* X *be a locally compact topological space and* A *a subspace of* M(X). *Then the following items are equivalent:*

(i) A *is relatively weakly compact;*

(ii) *given* $\varepsilon > 0$ *we have* A *satisfying the conditions*

(a) *there exists some compact* $K \subseteq X$ *such that*

$|\mu|(X\backslash K) < \varepsilon$ *for all* $\mu \in A$,

(b) *corresponding to any compact* $K \subseteq X$ *there is an open neighbourhood* O *of* K *such that*

$|\mu|(O\backslash K) < \varepsilon$ *for all* $\mu \in A$;

(iii) *there is a sequence* $\{\mu_n\}$ *in* A *such that, for all Borel* $E \subset X$ *such that* $\mu_n(E) = 0$ $(n \in \mathbb{N})$ *we have* $\mu(E) = 0$ $(\mu \in A)$.

B.5. NOTATIONS. Let A_1 and A_2 be normed linear spaces. A linear map T of A_1 into A_2 is said to be *weakly* (or *norm*) *compact* if it carries every bounded subset of A_1 into a weakly (or norm) compact subset of A_2.

The following Theorem is a special form of the Dunford-Pettis property - see e.g. Day [26, page 108] and Dunford and Schwartz [29] for the general cases.

B.6. THEOREM. *Let* X, Y *and* Z *be compact topological spaces and* $T_1 : M(X) \to M(Y)$ *and* $T_2 : M(Y) \to M(Z)$ *weakly compact linear maps. Then the composite linear map* $T_2 \circ T_1 : M(X) \to M(Z)$ *is norm compact.*

3. WEAKLY COMPACT SUBSETS OF C(X)

B.7. THEOREM (Grothendieck [54]). *Let* X *be a topological space and* A *a subspace of* $C(X)$. *Then the following items are equivalent:*

(i) A *is relatively weakly compact;*

(ii) *if* $\{f_n\}$ *and* $\{x_m\}$ *are sequences in* A *and* X, *respectively, then we have that the closures of the sets* $\{f_n(x_m) : n < m\}$ *and* $\{f_n(x_m) ; n > m\}$ *are not disjoint, (or equivalently whenever the double limits* $a := \lim_n \lim_m f_n(x_m)$ *and* $b := \lim_m \lim_n f_n(x_m)$ *exist, we must have* $a = b$).

B.8. NOTATION. Let X and Y be topological spaces and $P(X) := \{\nu \in M(X) : \nu \text{ is positive and } \|\nu\| = 1\}$. Corresponding to each function F in $C(X \times Y)$ we define the map $F' : P(X) \times P(Y) \to \mathbb{C}$ by

$$F'(\mu,\nu) := \iint F(x,y)\,d\nu(x)\,d\mu(y) \quad (\nu \in P(X) \text{ and } \mu \in P(Y)).$$

For all $x \in X$ and $\nu \in P(X)$ the functions $F(x,.)$ and $F'(\nu,.)$ defined on Y and P(Y), respectively, are given by

$$F(x,.)(y) := F(x,y) \quad \text{and} \quad F'(\nu,.)(\mu) := F'(\nu,\mu)$$

for all $y \in Y$ and $\mu \in P(Y)$. Similarly one defines the functions $F(.,y)$ and $F'(.,\mu)$ on X and P(X), respectively, for all $y \in Y$ and $\mu \in P(Y)$.

Evidently $F(x,.) \in C(Y)$ and $F(.,y) \in C(X)$ for all $x \in X$ and $y \in Y$. More generally we have

B.9. LEMMA. *Let* X *and* Y *be topological spaces and* F *and* F' *as defined above. Then*

(i) *The map* $x \to F'(\bar{x},\mu)$ *of* X *into* $\mathbb{C}$ *is continuous* ($\mu \in P(Y)$).

(ii) *If the set* $\{F(x,.) : x \in X\}$ *is relatively weakly compact, then the map* $x \to \phi(F(x,.))$ *of* X *into* $\mathbb{C}$ *is continuous* ($\phi \in C(Y)^*$).

Proof. (i) Given $\varepsilon > 0$, let K be a compact set such that $\mu(X\backslash K) < \frac{\varepsilon}{4a}$ where $a := \|F\|_{X\times Y}$. For each $x \in X$ we can find a neighbourhood $O(x)$ of x such that

$$|F(x,y) - F(t,y)| < \tfrac{1}{2}\varepsilon \quad \text{for all } t \in O(x) \text{ and } y \in K.$$

Hence, for all $t \in O(x)$, we have

$$\begin{aligned} |F'(\bar{x},\mu)-F'(\bar{t},\mu)| &= \left|\int (F(x,y) - F(t,y))d\mu(y)\right| \\ &\le \int_K |F(x,y) - F(t,y)|d\mu(y) + \int_{X\backslash K} |F(x,y) - F(t,y)|d\mu(y) \\ &\le \tfrac{1}{2}\mu(K)\varepsilon + 2\|F\|_{X\times Y}\,\mu(X\backslash K) \\ &< \varepsilon. \end{aligned}$$

(ii) Let (x_α) be a net converging to a point x in X. Since the net $(F(x_\alpha,.))$ is relatively weakly compact, it has a weak cluster point f in C(Y). For each $y \in Y$ we have

$$F(x_\alpha,.)(y) := F(x_\alpha,y) \to F(x,y) \quad \text{and so} \quad f(y) = F(x,y).$$

Thus $F(x,.) = f$ is the only weak cluster point of $(F(x_\alpha,.))$ and so our result follows.

B.10. THEOREM. *Let* X *and* Y *be topological spaces and* F *a function in* C(X × Y). *Recalling the notation in* B.8, *we have the following items equivalent:*

(i) *The sets* $\{F(x,.) : x \in X\}$ *and* $\{F(.,y) : y \in Y\}$ *are relatively weakly compact in* C(Y) *and* C(X), *respectively;*

(ii) *the sets* $\{F'(\nu,.) : \nu \in P(X)\}$ *and* $\{F'(.,\mu) : \mu \in P(Y)\}$ *are relatively weakly compact in* m(P(Y)) *and* m(P(X)), *respectively.*

Proof. That (ii) implies (i) is trivial. Now suppose (i) holds. For $\nu \in P(X)$. Let $F(\nu,.) \in C(Y)$ be the function given by $F(\nu,y) := F'(\nu,\bar{y})$ for all $y \in Y$. (Indeed $F(\nu,.) \in C(Y)$, by Lemma B.9(i).) Similarly define $F(.,\mu) \in C(X)$ for $\mu \in P(Y)$.

By Theorem B.2 we have the closure, call if $\tilde{\mathfrak{F}}_1$, of the convex hull of $\{F(\bar{x},.) : x \in X\}$ weakly compact. Fix $\nu \in P(X)$ with $K := \text{supp}(\nu)$ compact. We can find a set (λ_α) in the convex hull of $\{\bar{x} : x \in K\}$ such that $\lambda_\alpha(f) \to \nu(f)$ for all $f \in C(K)$. So, by Lemma B.9(ii), we have

$$\begin{aligned}\phi(F(\lambda_\alpha,.)) &= \int \phi(F(x,.))d\lambda_\alpha(x) \qquad \text{since } F' \text{ is linear}\\ &\to \int \phi(F(x,.))d\nu(x) \qquad (\phi \in C(Y)^*)\end{aligned}$$

But $F(\lambda_\alpha,y) = \int F(x,y)d\lambda_\alpha(x) \to \int F(x,y)d\nu(x) = F(\nu,y)$ $(y \in Y)$. Hence $\int \phi(F(x,.))d\nu(x) = \phi(F(\nu,.))$ $(\phi \in C(Y)^*)$. Thus $F(\nu,.) \in \tilde{\mathfrak{F}}_1$. By the inner regularity of measures we then have $F(\nu,.) \in \tilde{\mathfrak{F}}_1$, for all ν in P(X).

Hence $\{F(\nu,.) : \nu \in P(X)\}$ is a relatively weakly compact subset of C(Y). Recalling the symmetric criterion for weak compactness (see Theorem B.7), this is equivalent to saying $\{F'(.,\bar{y}) : y \in Y\}$ is a relatively weakly compact subset of m(P(X)). So the closure, say $\tilde{\mathfrak{F}}_2$, of the convex hull of $\{F'(.,\bar{y}) : y \in Y\}$ is weakly compact. By arguing in a similar manner to the preceding paragraph, we get that $F'(.,\mu) \in \tilde{\mathfrak{F}}_2$ for all $\mu \in P(Y)$. Consequently $\{F'(.,\mu) : \mu \in P(Y)\}$ and (by Grothendieck's Theorem B.7), $\{F'(\nu,.) : \nu \in P(X)\}$ are realtively weakly compact subsets of m(P(X)) and m(P(Y)), respectively.

Bibliography

1. R. Arens
   Operations induced in function spaces. *Monatsch. Math.* 55(1951), 1-19.
2. R. Arens:
   The adjoint of a bilinear operation. *Proc. Amer. Math. Soc.* 2(1951), 839-848.
3. W.G. Bade and H.G. Dales:
   Norms and ideals in radical convolution algebras. *Jour. Func. Analysis* 41(1981), 77-109.
4. A.C. Baker and J.W. Baker:
   Algebras of measures on a locally compact semigroup. *J. Lond. Math. Soc.* 1(1969), 249-259.
5. A.C. Baker and J.W. Baker:
   Algebras of measures on a locally compact semigroup II. *J. Lond. Math. Soc.* 2(1970), 651-659.
6. A.C. Baker and J.W. Baker:
   Algebras of measures on a locally compact semigroup III. *J. Lond. Math. Soc.* 4(1972), 685-695.
7. J.W. Baker and R.J. Butcher:
   The Stone-Čech compactification of a topological semigroup. *Math. Proc. Camb. Phil. Soc.* 80(1976), 103-107.
8. J.W. Baker and P. Milnes:
   The ideal structure of the Stone-Čech compactification of a group. *Math. Proc. Camb. Phil. Soc.* 82(1977), 401-409.
9. J.F. Berglund and K.H. Hofmann:
   Compact semitopological semigroups and weakly almost periodic functions. Springer Lect. Notes in Math. 42(1967).
10. J.F. Berglund, H.D. Junghenn and P. Milnes:
    Compact right topological semigroups and generalizations of almost periodicity. Springer Lect. Notes in Math. 663(1978).
11. F.F. Bonsall and J. Duncan:
    Complete normed algebras. Springer-Verlag, Berlin, New York (1973).

12. R.C. Buck:

Bounded continuous functions on a locally compact space. *Michigan Math. J.* 5(1958), 95-104.

13. R.B. Burckel:

Weakly almost periodic functions on semigroups. Gordon and Breach, New York (1970).

14. R.J. Butcher:

The Stone-Čech compactification of a topological semigroup and its algebra of measure. Ph.D. Thesis, University of Sheffield (1976).

15. Ching Chou:

On the size of the set of left invariant means on a semigroup. *Proc. Amer. Math. Soc.* 23(1969), 199-205.

16. Ching Chou:

On topologically invariant means on a locally compact group. *Trans. Amer. Math. Soc.* 151(1970), 443-456.

17. Ching Chou:

Topologically invariant means on the von Neumann algebra VN(G). *Trans. Amer. Math. Soc.* 273(1982), 207-229.

18. Ching Chou:

Weakly almost periodic functions and almost convergent functions on a group. *Trans. Amer. Math. Soc.* 206(1975), 175-200.

19. P. Civin and B. Yood:

The second conjugate space of a Banach algebra as an algebra. *Pac. J. Math.* 11(1961), 847-870.

20. A.H. Clifford and G.B. Preston:

The algebraic theory of Semigroups, Vol. l. Amer. Math. Soc. Providence, R.I. (1961).

21. P.J. Cohen:

Factorization in group algebras. *Duke Math. J.* 26(1959), 199-205.

22. W.W. Comfort and K.A. Ross:

Pseudocompactness and uniform continuity in topological groups. *Pac. J. Math.* 16(1966), 483-496.

23. I.G. Craw and N.J. Young:

Regularity of multiplication in weighted group and semigroup algebras. *Oxford Quart. J. Math.* 25(1974), 351-358.

24. H.W. Davis:
An elementary proof that Haar measurable almost periodic functions are continuous. *Pac. J. Math.* 21(1967), 241-248.

25. M.M. Day:
Amenable semigroups. *Ill. J. Math.* 1(1957), 509-544.

26. M.M. Day:
Normed linear spaces. Springer-Verlag, Berlin, New York (1973).

27. J. Diestal and J.J. Uhl Jr.:
Vector measures. Amer. Math. Soc. Providence, Rhode Island (1977).

28. J. Duncan and I. Namioka:
Amenability of inverse semigroups and their semigroup algebras. *Proc. Roy. Soc. Edinburgh* 80A(1978), 309-321.

29. N. Dunford and J. Schwartz:
Linear operators, Part I : General theory. Interscience, New York (1966).

30. H.A.M. Dzinotyiweyi:
Algebras of measures on C-distinguished topological semigroups. *Math. Proc. Camb. Phil. Soc.* 84(1978), 323-336.

31. H.A.M. Dzinotyiweyi:
Algebras of functions and invariant means on semigroups. *Math. Instit. Katholieke Universiteit, The Netherlands, Report* 7807 (1978).

32. H.A.M. Dzinotyiweyi:
On the analogue of the group algebra for locally compact semigroups. *J. Lond. Math. Soc.* 17(1978), 489-506.

33. H.A.M. Dzinotyiweyi:
Continuity of semigroup actions on normed linear spaces. *Oxford Quart. J. Math.* (2) 31(1980), 415-421.

34. H.A.M. Dzinotyiweyi:
Norm and weak continuity of semigroup actions on normed linear spaces. *Oxford Quart. J. Math.* (2) 33(1982), 85-90.

35. H.A.M. Dzinotyiweyi:
Weakly almost periodic functions and the irregularity of multiplication in semigroup algebras. *Math. Scand.* 46(1980), 153-172.

36. H.A.M. Dzinotyiweyi:
Compactness criteria for semigroups: a note on a preceding paper. *Math. Scand.* 46(1980), 173-176.

37. H.A.M. Dzinotyiweyi:
Nonseparability of quotient spaces of function algebras on topological Spaces. *Trans. Amer. Math. Soc.* 272(1982), 223-235.

38. H.A.M. Dzinotyiweyi and P. Milnes:
Functions with separable orbits on foundation semigroups. *J. Nigerian Math. Soc.* 1(1982), 31-38.

39. H.A.M. Dzinotyiweyi and G.L.G. Sleijpen:
A note on measures on foundation semigroups with weakly compact orbits. *Pac. J. Math.* 81(1980), 61-69.

40. H.A.M. Dzinotyiweyi and A.C.M. van Rooij:
Non-Archimedean harmonic analysis on topological semigroups. *Indagationes Mathematicae* 45(1983), 301-314.

41. W.F. Eberlein:
Abstract ergodic theorems and weak almost periodic functions. *Trans. Amer. Math. Soc.* 67(1949), 217-240.

42. R. Ellis:
Locally compact transformation groups. *Duke Math. J.* 24(1957), 119-125.

43. R. Giles:
A generalization of the strict topology. *Trans. Amer. Math. Soc.* 161(1971), 467-474.

44. L. Gillman and M. Jerison:
Rings of continuous functions. Van Nostrand, New York (1960).

45. I. Glicksberg:
Some remarks on absolute continuity on groups. *Proc. Amer. Math. Soc.* 40(1973), 135-139.

46. I. Glicksberg:
Weak compactness and separate continuity. *Pac. J. Math.* 11(1961), 205-214.

47. A.M.M. Gommers:
Non-Archimedean harmonic analysis on groups without Haar measure. Ph.D. Thesis, Katholieke Universiteit, The Netherlands (1979).

48. C. Gowrisankaran:
Radon measures on groups. *Proc. Amer. Math. Soc.* 35(1972), 503-506.

49. E.E. Granirer:
Exposed points of convex sets and weak sequential convergence. *Mem. Amer. Math. Soc.* 123(1972).

50. E.E. Granirer:
Weakly almost periodic and uniformly continuous functionals on the Fourier algebra of any locally compact group. *Trans. Amer. Math. Soc.* 189(1974), 371-382.

51. E.E. Granirer:
On amenable semigroups with finite dimensional set of invariant means, I and II. *Ill. J. Math.* 7(1963), 32-58.

52. E.E. Granirer:
The radical of $L^{\infty}(G)^*$. *Proc. Amer. Math. Soc.* 41(1973), 321-324.

53. F.P. Greenleaf:
Invariant means on topological groups. Van Nostrand, New York (1969).

54. A. Grothendieck:
Critères de compacité dans les espaces fonctionnels généraux. *Amer. J. Math.* 75(1952), 168-186.

55. A. Grothendieck:
Sur les applications linéaires faiblement compactes d'espaces du type C(K). *Canad. J. Math.* 5(1953), 129-173.

56. S.L. Gulick:
Commutativity and ideals in the biduals of topological algebras. *Pac. J. Math.* 18(1966), 121-137.

57. P. Halmos:
Measure theory. D. van Nostrand, Princeton (1950).

58. G. Hart:
Absolutely continuous measures on semigroups. Ph.D. Thesis, Kansas State University (1970).

59. E. Hewitt and K.A. Ross:
Abstract harmonic analysis, vol. I. Springer-Verlag, Berlin (1963).

60. E. Hewitt and H.S. Zuckermann:
The ℓ_1-algebra of a commutative semigroup. *Trans. Amer. Math. Soc.* 83(1956), 70-97.

61. R. Hirschfeld:
Riots. *Niew Arch. Wisk.* 22(1974), 1-43.

62. K.H. Hofmann and P.S. Mostert:
Elements of compact semigroups. Merill, Columbus, Ohio (1966).

63. J.W. Jenkins:
Amenable subsemigroups of a locally compact group. *Proc. Amer. Math. Soc.* 25(1970), 766-770.

64. B.E. Johnson:
Cohomology in Banach algebras. *Mem. Amer. Math. Soc.* 127(1972).

65. J.L. Kelley:
General topology. Princeton, Van Nostrand (1955).

66. J.L. Kelley and I. Namioka:
Linear topological spaces. D. van Nostrand, Princeton, New Jersey, London (1963).

67. F. Kinzl:
Absolut stetige Masse auf lokalkompakten Halbgruppen. *Monatshefte für Math.* 87(1979), 109-121.

68. M.M. Klawe:
Dimensions of the sets of invariant means of semigroups. *Ill. J. Math.* 24(1980), 233-243.

69. C.D. Lahr:
Approximate identities for convolution measure algebras. *Pac. J. Maths.* 47(1973), 147-159.

70. R. Larsen:
Transformation groups and measures with separable orbits. *Monatsh. Math.* 73(1969), 222-225.

71. A.T. Lau:
Invariant means on subsemigroups of locally compact groups. *Rocky Mountain J. Math.* 3(1973), 77-81.

72. M. Leinert:
A commutative Banach algebra which factorizes but has no approximate units. *Proc. Amer. Math. Soc.* 55(1976), 345.

73. T.S. Liu, A. van Rooij and J.K. Wang:
Group representations in Banach spaces: Orbits and almost periodicity. *Studies and essays presented to Yu-Why Chen on his sixtieth birthday* (1970), 243-254.

74. I. Namioka:
On certain actions of semigroups on L-space. *Studia Math.* 29(1967), 63-77.

75. A. Olubummo:
Finitely additive measures on the non-negative integers. *Math. Scandinavica* 24(1969), 186-194.

76. A. Olubummo:
A decomposition theorem for finitely additive measures on a discrete commutative semigroup. *Nigerian J. Science* 4(1970), 101-110.

77. A.B. Paalman-de-Miranda:
Topological semigroups. Math. Centre Tracts, Amsterdam (1970).

78. W.L. Paschke:
A factorable Banach algebra without bounded approximate unit. *Pac. J. Math.* 46(1973), 249-251.

79. A.L.T. Paterson:
Invariant measure semigroups. *Proc. Lond. Math. Soc.* 35(1977), 313-332.

80. A.L.T. Paterson:
Amenability and locally compact semigroups. *Math. Scand.* 42(1978), 271-288.

81. J.S. Pym:
Convolution algebras are not usually Arens-Regular. *Quart. J. Math Oxford* (2) 25(1974), 235-240.

82. J.S. Pym:
The convolution of functionals on spaces of bounded functions. *Proc. Lond. Math. Soc.* 15(1965), 84-104.

83. J.S. Pym:
Weakly separately continuous measure algebras. *Math. Annalen* 175 (1968), 207-219.

84. J.S. Pym and H.L. Vasudema:
An algebra of finitely additive measures. *Studia Math.* 54(1975), 29-40.

85. J.S. Pym and H.L. Vasudema:
The algebra of finitely additive measures on a partially ordered semigroup. *Studia Math.* LXI (1977), 7-19.

86. Roger Rigelhof:
Invariant measures on locally compact semigroups. *Proc. Amer. Math. Soc.* 28(1971), 173-176.

87. N.J. Rothman:
An L^1-algebra for certain locally compact topological semigroups. *Pac. J. Math.* 23(1967), 143-151.

88. W. Ruppert:
On semigroup compactifications of topological groups. *Proc. Roy. Irish Academy* 79A(1979), 179-200.

89. H.H. Schaefer:
Banach lattices and positive operators. Springer-Verlag, Berlin, New York (1974).

90. F.D. Sentilles:
Compactness and convergence in the space of measures. *Ill. J. Math.* 13(1969), 761-768.

91. S.H. Simmons:
A converse Steinhaus theorem for locally compact groups. *Proc. Amer. Math. Soc.* 49(1975), 383-386.

92. G.L.G. Sleijpen:
Convolution measure algebras on semigroups. Ph.D. Thesis, Katholieke Universiteit, The Netherlands (1976).

93. G.L.G. Sleijpen:
Emaciated sets and measures with continuous translations. *Proc. Lond. Math. Soc.* 37(1978), 98-119.

94. G.L.G. Sleijpen:
Locally compact semigroups and continuous translations of measures. *Proc. Lond. Math. Soc.* 37(1978), 75-97.

95. G.L.G. Sleijpen:
The support of the Wiener algebra on stips. *Indagationes Math.* 42(1980), 61-82.

96. G.L.G. Sleijpen:
The action of a semigroup on a space of bounded Radon measures. *Semigroup Forum.* 23(1981), 137-151.

97. G.L.G. Sleijpen:
Norm- and weak-continuity of translations of measures; a counter-example. *Semigroup Forum* 15(1978), 343-350.

98. G.L. G. Sleijpen:
A note on measures vanishing on emaciated sets. *Bull. Lond. Math. Soc.* 13(1981), 219-223.

99. J.D. Stein:
Uniform absolute continuity in spaces of set functions. *Proc. Amer. Math. Soc.* 51(1975), 137-140.

100. K.W. Tam:
On measures with separable orbit. *Proc. Amer. Math. Soc.* 23(1969), 409-411.

101. J.L. Taylor:
The structure of convolution measure algebras. *Trans. Amer. Math. Soc.* 119(1965), 150-166.

102. A.C.M. van Rooij:
Non-Archimedean functional analysis. Marcel Dekker. Inc., New York and Bassel (1978).

103. A.C.M. van Rooij:
Tight functionals and the strict topology. *Kyungpook Math. J.* 7(1967), 41-43.

104. J.C.S. Wong:
Abstract harmonic analysis of generalized functions on locally compact semigroups with applications to invariant means. *J. Austral. Math. Soc.* 23(1977), 84-94.

105. N.J. Young:
The irregularity of multiplication in group algebras. *Quart. J. Math. Oxford* 24(1973), 59-62.

106. N.J. Young:
Semigroup algebras having regular multiplication. *Studia Math.* 47(1973), 191-196.

Index of symbols

$M_n(S)^\wedge$	non-zero multiplicative linear functionals on $M_n(S)$	166
$M^{\ell}_{eq-a}(S)$, $M^{r}_{eq-a}(S)$, $M_{eq-a}(S)$	sets of equi-absolutely continuous measures on S	117
$M^{\ell}_{eq}(S,H)$	left equi-quasi-invariant measures	35
$M^{\ell}_{q}(S)$	left quasi-invariant measures	35
$m(X)$	space of bounded complex-valued functions on X	3
$\|\mu\|$	measure arising from total variation of μ	4
$\|\mu\|_*$	inner measure arising from $\|\mu\|$	4
$\bar{\mu}$	the completion of a measure μ	4
$\mu\|_B$	restriction of a measure μ to set B	4
$\hat{\mu}$	the Fourier transformation $\mu \to \hat{\mu}$ defined on M(S)	40
$\mu_1 * \mu_2$	the convolution of two measures μ_1 and μ_2	7
$N_\mu(x)$	$\inf\{\|\mu\|(O) : O$ an open neighbourhood of $x\}$	155
$\Vert\ \Vert$	norm (e.g. total variation norm on M(X))	4
$\Vert\ \Vert_X$	supremum norm on space of functions on X	3
$o(G)$	$\{n \in \mathbb{N} : G/H$ has an element of order n, for some open subgroup $H\}$	155
P_ψ	a pseudo-norm on m(X) arising from $\psi \in m(X)$	174
$P(H)$ - LIM	a mean left invariant with respect to $P(H)$	72
S	given (topological) semigroup [in most cases]	1
S_1	smallest dense ideal of a stip S	46
S_r	the set $\{x \in S :$ for all $f \in C_K(X,Y)$, $f_x \in C(S,Y)$ for any topological space $Y\}$	17
$S^\wedge$	continuous semicharacters of S	38
$S^\wedge_n$	non-Archimedean continuous semicharacters that are not $M_n(S)$-negligible	166
supp(μ)	the support of a measure μ	4

TLIM	topological left invariant mean	72
T_K	roots of unity not divisible by characteristic of k	153
UC(S)	uniformly continuous functions on S	63
UC(S^)	non-Archimedean uniformly continuous functions on S^	163
$\mathcal{U}_T$	non-Archimedean uniformity induced by a compact set $T \subseteq S$	163
WAP(S)	weakly almost periodic functions on S	63
WUC(S)	weakly uniformly continuous functions on S	63
$\bar{x}$	point mass at x for $x \in S$, or the natural image of $x \in K$ in k.	4
χ_E	characteristic function of a set E	4
χ_o-clo(F)	countable closure of a set F	53

Subject index